Developments in Landscape Management and Urban Planning, 6D

New challenges in recreation and tourism planning

FURTHER TITLES IN THIS SERIES

ISOMUL

ISOMUL = International Studygroup on Multiple Use of Land. This studygroup encourages, guides and publishes studies in the field of the multiple uses of land for rural and urbanized areas. Members are found in several European and North-American countries.

Developments in Landscape Management and Urban Planning, 6D

New challenges in recreation and tourism planning

Edited by

Hubert N. van Lier
Department of Rural Planning and Development, Agricultural University,
6703 BJ Wageningen, The Netherlands

Pat D. Taylor
Program in Landscape Architecture, The University of Texas at Arlington, Arlington,
TX 76019-0108, U.S.A.

ELSEVIER
Amsterdam — London — New York — Tokyo   1993

ELSEVIER SCIENCE PUBLISHERS B.V.
Sara Burgerhartstraat 25
P.O. Box 211, 1000 AE Amsterdam, The Netherlands

ISBN: 0-444-89849-2

Printed in The Netherlands

FOREWORD

This fourth ISOMUL book puts forth new challenges and new methods for the planning of outdoor recreation and tourism. Chapters 1 and 2 offer a description of the problems which the public and the private sectors are facing. Following chapters deal with several aspects of the problems, including trends and changes in outdoor recreation and tourism, previous policies of planning for recreation and tourism, shortcomings in policies and actions, and new approaches in several countries, such as the United States, France, Switzerland and the Netherlands. Finally, the book gives examples of new methods and approaches for the planning for outdoor recreation and tourism. These methods are demonstrated with specific areas or projects from the afore mentioned countries.

ISOMUL, the International Studygroup on the Multiple Use of Land, was formed by a small group of scientists from various nations, all having an interest in the field of rural land-use planning. It seeks to establish an exchange of information and experience by encouraging and guiding studies in land-use planning. Because of a lag in the application of research, there is a need to transfer knowledge and understanding about planning for future uses of land in rural areas, and particularly rural lands which are most susceptible to dynamic forces such as socio-economic development, technology, and metropolitan influences. Each country, dealing with its own problems, has specific knowledge which is valuable to others. The exchange of this type of information is a necessary and valuable activity and is the major objective of ISOMUL. It, therefore, encourages studies which focus on specific planning problems such as multiple land-use planning methods, forecasting and prediction methods, evaluation methods and impact analyses.

The first book of ISOMUL, written by R. Burnell Held (Colorado State University, U.S.A.) and D.W. Visser (Governmental Service for Land and Water Use, The Netherlands) dealt with a comparison of land-use developments and land-use planning policies between the two countries. The second multi-authored book, edited by F.R. Steiner (Washington State University, U.S.A.) and H.N. van Lier (Agricultural University, Wageningen, The Netherlands) dealt with the basic problems and approaches to conservation and development of land resources. The third ISOMUL book, edited by W.W. Budd (Washington State University, U.S.A.), I. Duchhart (Agricultural University, Wageningen, The Netherlands), L.H. Hardesty (Washington State University, U.S.A) and F. R. Steiner (Arizona State University, U.S.A.) deals with agroforestry, a system which tries to overcome serious social and environmental problems. Although a large variety of agroforestry systems exist, they all are integrated within the rural land uses and especially within farming methods. The agroforestry book describes several planning methods for implementing agroforestry systems.

It is hoped that this fourth ISOMUL book, "New Challenges in Recreation and Tourism Planning", will help all those who bear the responsibility of providing human beings with possibilities and facilities to spend their free time in their own desired way.

Wageningen   Prof.dr.ir. H.N. van Lier
October 1992   Chairman, ISOMUL

CONTENTS

CHAPTER 1

Preface

by
Pat D. Taylor
President, Taylor & Associates, Dallas, USA
Director, Program in Landscape Architecture
The University of Texas at Arlington, U.S.A.
and
Adjunct Associate Professor, Department of Recreation, Park and Tourism Sciences
Texas A&M University, U.S.A.

The world is always changing, or so the adage goes. But, as if to provide balance to the universe, there exists a counter adage: There is nothing new, only ignorance revisited.

Each adage is rooted in truth, depending upon the perspectives of the speaker and the times which he or she is describing. While today's world is witness to breathtaking alterations in political and economic order, and while scientific knowledge leaps forward with advances faster than the information age can absorb them, these developments are used by some to call out new "truths" about the human condition, as if the evidence has never before been seen.

The world of recreation and tourism is astir today, just as vigorously as are the worlds of politics, economics and science. Neophyte observers of new truths fervently believe that they have discovered a panacea for the new world disorder: that recreation and tourism are best suited to unite humankind into the cohesive common purpose of making the world's dynamic social changes permanent.

Others--those with contrary views--caution against such euphoria. They remind us that the general pattern of the world is evolutionary and not revolutionary. Thus, we should be hesitant to put so many recreation and tourism eggs into the world's porous social basket.

The truth, no doubt, is somewhere in between. Recreation and tourism, along with other human endeavors, can either contribute to social stability or be disruptive to it, depending on the circumstances. In some cases they will do both, as in those societies where foreign investment is needed to promote public infrastructure, yet where the initial attractions will be altered with increased visitation.

The scientists, practitioners and policy makers who have contributed to this text, have delivered several seasonings to this olio. Their approaches and perspectives about the efficacy of recreation and tourism are as varied as their international origins.

From Switzerland, Dr. Willy Schmid, along with Dr. Janos Jacsman and René Schilter, outline new techniques for reacting to tourism's social, cultural and physical impacts. These Swiss scientists share the view of Dr. Carson Watt, Dr. Turgut Var and Dr. James Stribling of Texas A&M University that the costs of tourism must be assessed at the same time we project its benefits.

Although costs are not a part of their particular chapter, the Texas A&M scientists provide useful hands-on techniques for assessing needs and developing goals at the local tourism level.

A global view is offered by Dr. Donald F. Holecek of Michigan State University. His view of the future suggests that world demand during the next decade will outstrip supply of recreation and tourism opportunity.

Dr. Jean Tarlet convincingly presents the French experience in tourism and recreation as a prototype with global implications. In addition to tracing the historical evolution of France's leisure industries, Tarlet demonstrates with site-specific examples his country's response to consumer demand. He also reports on the impact of recreation and tourism on sensitive social and physiographic settings.

Dr. Adri Dietvorst articulately explains the evolution of Dutch involvement in complex planning processes, and then demonstrates Dutch adaptiveness through the application of market oriented approaches to public processes once closed to such economic strategies. Dr. Hubert N. van Lier, in his introduction, echoes the deepening appearance of private sector strategies in the ever-widening arena of public sector decision making.

Mr. Hubertus J. Mittmann stresses the resource management side of recreation and tourism. His report on multiple use management within lands owned by the United States Forest Service reminds us that despite the co-mingling of private and public strategies, the costs of public ownership remain perpetual for government agencies. To add to the weighty responsibility, Mittmann reminds us further that after the consumer has paid market price for a recreation opportunity, he or she walks away with a value of that opportunity which exceeds the cost, and to Mittmann, this is appropriate justification for sustained management by the public sector.

Mr. Joe Porter provides testimony to the utility of the planning and design processes. His call for increased dependence by clients, public and private, on these processes is coupled with verification of success through his consulting work in resort communities.

Dr. David L. Edgell adds to the call for vigilance in protecting local customs and cultures from the costs which accompany tourism. In addition, he outlines the timely, intriguing, and pressing issues behind political and foreign policy implications of international tourism.

Timeliness reminds us of trends, which in turn remind us to keep an eye on the middle ground. In a sense the recent embrace by public officials, including educators, of so many private sector principles has added "real world" credibility to the language, thoughts and practices of people often thought of as out-of-touch. Many of us reveled in our new found abilities to think like entrepreneurs, when in fact we have only rediscovered truths long practiced in private sector circles.

The cost of these new truths, for many, has been the abandonment of some other truths regarding the roles of government and public policy in the delivery of recreation and tourism opportunity. During the last decade in America, innovations in recreation and tourism have been driven by product manufacturers, skilled at developing new items for recreational use, resulting in new types of recreation behavior. From athletic wear to mind games, from devices which fly, skim or propel people through the air, over the ground or across the water, to the latest in virtual reality electronic games, product manufacturers have read America's recent flirtation with market forces, and have enjoyed the return which comes from deep pockets of investment capital.

Are product manufacturers, then, to be the harbinger of Europe's rediscovery of market systems? Are we to assume that the most the world will see from the loosening-up of governmental controls, the elimination of social and economic subsidies, and the growth of private wealth in eastern and western Europe, will be a decade of gadgets?

No doubt, such an approach to recreation and leisure would be fun and exciting, but it would be apt to come at considerable public cost, if America's experience of the 1980's is any lesson. While we have adopted countless new ways to recreate, the great public spaces on which we rely to do it have fallen into disrepair. From highways and bridges which take us to our playgrounds, to the quality of the earth, water and air resources which sustain us, we have failed to care in meaningful ways. While it is of little comfort, this disparity between consumer based satisfaction and care for our common resources has always been hard to balance in

America. We've seen these cycles before. This is not the first time either privatization or subsidy have fallen from grace or come into fashion. What we must remember about fashion cycles is what the designers of the innovatively conservative, yet popular Volvo advocate: What is fashionable today is unfashionable tomorrow.

And, so it goes with recreation and tourism, and so it goes with sound public policy. While market application can improve the delivery of some public services, not every undertaking in recreation and tourism can be privatized. Not every undertaking should be privatized.

Again, this situation is nothing new, it's just ignorance revisited. Anytime that a balanced situation--such as that between public need and private gain, indigenous to mixed economies--is tilted, we wind-up justifying our not carrying-out this or that public service because ostensibly the money is not there.

It's not that money is lacking for legitimate projects, public or private. What lacks is commitment. When the commitment to innovate is there, money follows. The private sector understands this. Investors always will find capital for worthwhile projects. Despite trends and despite difficult political repercussions, it's a lesson that those of us who support public policy must never forget.

This book originated under the auspices of ISOMUL (the International Studygroup for the Multiple Use of Land) and was forged primarily from papers presented at the 1990 conference of CELA (the Council of Educators in Landscape Architecture) in Denver, Colorado, USA. The editors wish to thank ISOMUL and CELA, along with the agencies, corporations and universities with which the contributing authors are associated.

CHAPTER 2

New challenges in recreation and tourism planning

Introduction

by
Hubert N. van Lier
Professor and Chairman
Department of Physical Planning and Rural Development
Head Section Land and Water Design
Agricultural University, Wageningen, The Netherlands

2.1   Recreation and tourism on the move

Moving--or more precisely, mobility--is essential for tourists, recreationists, and vacationers. It always has been, and it always will be. Even though tourists are likely to spend time in one or two locations, the ability to move is essential since these leisure undertakings usually take place away from home.

While mobility made tourism possible, the industry today faces all the choices associated with complex socio-economic issues. Among these is the reality that demand is rapidly changing for the activities performed on these leisure retreats. Compared with a decade ago, human behavior, lifestyles and consumer demands show significant variances. Several of the authors in this text (Mittmann, Schmid et al., Dietvorst and Tarlet), for example, have observed that there is a tendency toward more but shorter periods for recreation--toward recreation on an individual basis or in small groups, and towards more active forms of outdoor recreation.

These dynamic developments in demand have repercussions on the supply side of the balance sheet, as well. Facilities must be adjusted to the changing tastes and preferences of the recreation and tourism consumer. These changes in consumer behavior are influenced by a number of factors including increasing awareness of environmental problems, the revenue-generating objectives of sustainable developments and of product manufacturers, and the pressure to reduce surplus production in rural agricultural lands, to name a few. In many countries these phenomena have led to the development of multiple-use policies in which outdoor recreation is combined with other critical land uses: farming, nature protection, or even urban and industrial development.

Another changing aspect of the industry involves the management of facilities. From the 1950's through the 1970's, many facilities were provided by governmental bodies, especially in western Europe. Today, the impossibility of these bodies to transform the facilities according to changing demands, and to manage them accordingly, has made it necessary to look for other approaches, such as privatizing.

Whereas in former days the planning of outdoor recreation facilities focused on large scale projects somewhere in the countryside and serving many demand centers, the need today is for smaller provisions close-in, where demand drives the type of facilities, services and activities offered.

2.2   Planning, Design and Management

An important aspect of planning for tourism and recreation is the acceptance of uncertainties in the planning and design processes. Planning is no longer seen as the result of a linear process along such lines as determination of demand (amount), the location of facilities, or their design, execution and use. It is accepted that as society changes, facilities must change to meet new demands. This fact alone necessitates the need for richer, more market-oriented planning.

The same new thinking applies for facility management. Many older facilities no longer provide the needs of tourists, and such facilities have become expensive to manage and maintain. This reality is particularly true for public projects where private input has been involved minimally. The creation of responsive outdoor recreation facilities, we have learned, requires a combination of public and private participation, if not in their management then certainly in their planning.

In many western European countries there was a boom in the planning of public outdoor recreation facilities in the 1960's and 1970's. Both national as well as regional and local authorities developed policies to plan and implement outdoor recreation, independent of the activities of the private sector. The result was a great increase in facilities such as swimming pools, sports fields, public beaches (on seashores and inland lakes), trails, picnic sites, forest lands, public campsites, and the opening-up of waterbodies to marinas and their developments. At the same time the private sector also grew active with hotels, campsites, specialized playgrounds and investment-intensive theme parks. It was a boom period.

However, in the 1980's, the limits were met, especially those for public agencies, in their ability to facilitate tourism and recreation. The most important limits included:

*   The high management and maintenance costs of older public sites, which cut into the delivery of adequate services. For example, scores of local

swimming pools had to be closed because of lack of money to manage and maintain them.

* The changing of recreation behaviors, causing the need for new and different facilities or the reconstruction of older ones. One example was the demand for shorter but more frequent swimming opportunities, which required more covered pools. Public providers found, in many cases, there was either no money available, or there was a lack of political willingness to find more money.

* The economic crisis of the 1980's had as one of its consequences decreasing budgets at the national, regional and local levels, while at the same time there was an increasing need for unemployment money. This competition for public money resulted in smaller funds for, among other things, outdoor recreation. While there were many plans, there were fewer public funds, resulting in fewer new provisions.

To solve this problem, it became increasingly popular to ask what needs could be provided by the private sector rather than by traditional public sources. This in turn raised such questions as:

* Are tourism and recreation economic activities?
* Is it possible to "economize" tourism and recreation?

Recreation and Tourism: Economic Activities?
The philosophy behind public involvement in the period between 1960 and the late 1970's was that recreation was needed to recover from the grind of daily work. It was a social need most of all. Therefore, it was appropriate that the public sector be held responsible for providing recreation opportunity for everyone, in addition to what the private sector was doing. Figure 2.1 summarizes those historically involved in outdoor recreation delivery.

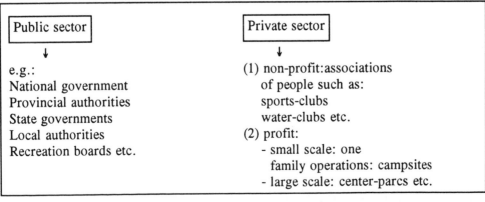

Figure 2.1   THE DIFFERENT INTERESTS-GROUPS IN OUTDOOR-RECREATION PLANNING AND POLICIES

Because of the strong increase in facilities and activities during former decades, recreation and tourism were increasingly perceived as economic activities. This not only meant that benefits were expressed in terms of money but also that management and maintenance, along with planning and policy making were studied in economic terms: How much does it cost, how can costs be reduced? Figure 2.2 gives the changes in conditions:

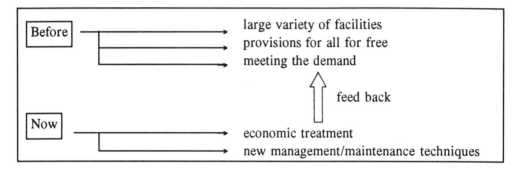

Figure 2.2   THE CHANGING ROLE OF PUBLIC BODIES IN OUTDOOR RECREATION POLICY MAKING AND PLANNING

In contrast, some figures can be given to demonstrate the economic importance of tourism and recreation at this moment. Figure 2.3 gives the role that tourism and recreation played in the E.C. in 1988.

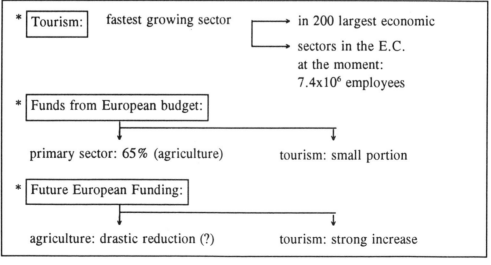

Figure 2.3   THE ROLE OF TOURISM AND RECREATION IN EUROPE (VAN LIER, 1989)

In the Netherlands tourism and recreation have a 'production' value of $15x10^9$ dollars with $240x10^3$ employees. In comparison: agriculture (the Netherlands is the second largest food exporter in the World) produced $11.5x10^9$ dollars with $260x10^3$ employees. It is foreseen that in the future there will be a strong increase in tourism and recreation while farming will loose pace.

The money spent on outdoor recreation in the Netherlands is $8x10^6$ dollars. The spending per person per day on outdoor recreation sites varies from 4 to 12 dollars. The number of visitors to 347 specific sites for more extensive forms of outdoor recreation, with an average size of 160 acres, totaled $44x10^6$ persons (equaling $127x10^3$ persons per project per year). For 46 intensively used sites, with an average size of 100 acres, these figures totaled $17x10^6$ persons (or $374x10^3$ persons per project) per year. Vacation recreation was done by 73.5% of the people, together 'producing' $24x10^6$ vacation-trips of which $14x10^6$ were in the Netherlands and $10x10^6$ were abroad. The spending of this last group is equal to the selling of natural gas abroad, or 2 times as high as the export of Dutch dairy products. This causes a negative 'tourism-balance' of $3.8x10^9$ dollars, which is likely to grow to $4x10^9$ or more in the future.

To economize tourism and recreation
As stated, the need to 'economize' tourism and recreation is a consequence of cutting down on public budgets. The public sector, under these new conditions, has had to search for new options, such as:

* accepting a more economic approach to outdoor recreation;
  balancing benefits and costs with an emphasis on reducing costs;
* introducing modern forms of management, product definition, and marketing
  (Jacsman et al., 1990; Mittmann, 1990; and Tarlet, 1990).

Generally, a more economic approach for the public sector to provide public goods can be accomplished in three ways (Koopmans and Wellink, 1987):

(1)   By increasing the efficiency of delivering their goods;
(2)   By introducing different allocation systems in order to deliver provisions
        closer to demand. Such possibilities are:
        . using the profit-principle
        . privatizing
        . decentralizing
        . deregulation;
(3)   By decreasing the level of provision.

Another term frequently used is commercializing, which can be considered as a combination of privatizing and applying the profit principle with one clear objective: To make a profit. Figure 2.4 demonstrates this approach.

10

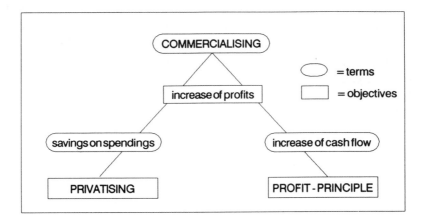

Figure 2.4   COMMERCIALIZING AS RELATED TO PRIVATIZING AND
THE PROFIT PRINCIPLE (AFTER VAN GELDER, 1987).

Privatizing is an instrument, used by many governments in the E.C. to reduce
the amount of collective spending. It can be defined as 'all changes in the provision
of a service or good in which the private sector increases relative to the public
sector until a point in which the public sector moves out completely' (Edwards,
1983). Figure 2.5 demonstrates this approach.

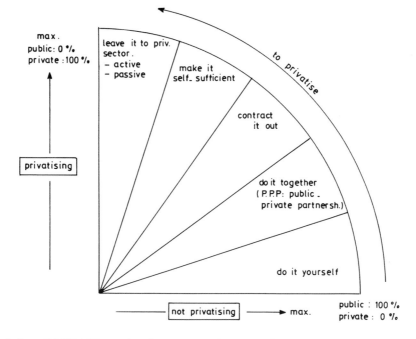

Figure 2.5   PRIVATIZING OF PUBLIC RESPONSIBILITIES (VAN DAM,
1988)

As the figure shows, many forms of privatizing exist, including:

* managing by contract: the public sector retains the ultimate responsibility, while the private sector does the exploitation.
* leaving it partially to the private sector: the public sector, however, may be actively involved in setting policies and in stimulating the private sector. For example, the management and maintenance of a facility can be done by contract, while the public sector can retain ownership of the property of development.
* leaving it completely to the private sector: thus, it is left to the free market (Philipsen, 1988).

Several motives are used for privatizing, most important among them being:

1. the motive of better management by a private organization: because of higher efficiency, the private sector provides the same good at lower prices, based upon three reasons:
   * a private organization creates an optimal size,
   * competition presses an increase in efficiency; and,
   * management is highly motivated because of its direct interest in the results.
2. a budget motive: briefly it aims to reduce public spending (see Koopmans and Wellink, 1987, and Kroese and Slangen, 1986).
3. a decision motive: the public sector has become so complicated that quick and efficient decisions are hard to make.
4. a social motive: decision making and planning are left to the direct interest-groups such as clubs, foundations or user groups.

The profit principle can be defined as making the user pay part or all of the cost or market price of a given public good or service. There are at least three motives for adapting the profit principle:

1. The fair distribution motive: some users pay too much and others pay too little for the public good, compared to their use.
2. The allocation motive: the profit principle forces each individual to make a decision of use versus non-use based upon the costs he or she incurs.
3. The budget motive: the budget problems of the public sector can be eased through the increase of income.

Based upon the analysis of privatizing and the profit principle, the term economizing can be defined as all changes in the provision of a good or service, for which the objective is to maximize the social and/or financial benefits, through special activities related to cost decreases and/or cash flow increases. The next figure illustrates the principle of economizing.

12

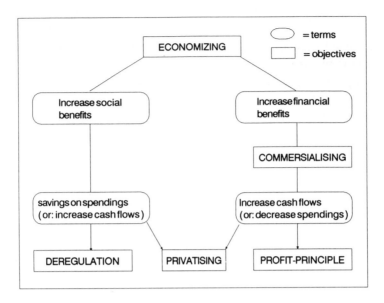

Figure 2.6   ECONOMIZING, AS DEFINED IN ITS TERMS, OBJECTIVES
AND MEANS

Given the splitting responsibilities between public and private sectors, an effort
has taken place to privatize recreation in the Netherlands, as the following figure
shows.

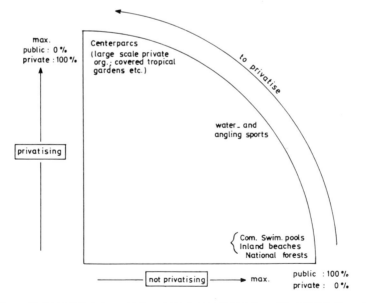

Figure 2.7   EXAMPLES   OF   PRIVATIZING   SPECIFIC   OUTDOOR
RECREATION FACILITIES IN THE NETHERLANDS

As noted, the public sector is being forced to look for a better balance between its responsibilities and those of the private sector. However, apart from the many forms which this balance can take there are at least two additional problems.

First, many regions or larger sites have components some of which can and some of which cannot be privatized. Not only is it a large problem to distinguish these components, but it is equally difficult to find the right methods to conduct the search for them.

In addition, the public sector has responsibilities other than providing outdoor recreation. Privatizing can have negative effects on these other responsibilities. In the Netherlands, for example, the recreational use of the countryside is considered carefully by the public sector because it occurs where other activities take place which are vital to the public interest.

2.3   Conclusions

As stated, the behaviors, policies and approaches to recreation and tourism change. As several authors observe in this text, these changes can be expressed in terms of frequencies, time spent, activities and group processes (see Dietvorst, Tarlet, Mittmann; Watt, Var and Stribling; and Jacsman, Schilter and Schmid). Changing attitudes, behaviors, needs and demand require different approaches in planning and designing facilities for recreation and tourism (see Porter). Most of the authors in this text focus specifically on one of several approaches, such as cyclic planning methods, or as Dietvorst calls it, stimulative planning. Mittmann describes multiple land-use planning, while Tarlet talks about product orientation, and Schmid et al. describe new concepts for specific sites.

Watt, Var and Stribling, along with Schmid, Jacsman and Schilter, promote a method for diffusing market oriented practices to private (as well as public) providers. In addition, they stress the need to balance the benefit-cost analysis of recreation and tourism with more emphasis on costs.

Tarlet, along with Schmid, Jacsman and Schilter, also point out that the manage-ment and maintenance aspects of recreation and tourism should be given more attention. They stress the economizing of tourism and recreation, as well as the fact that both public and private involvement play an important role.

Planning and design methods show new developments and approaches as well. Planning and design teams demand special management techniques as demonstrated by Porter in his report on resort community development. Special attention is given to the planning and design processes themselves, all aiming at environmental quality.

Holecek evaluates trends in world social and economic activity, and their impact on recreation and tourism, while Edgell expands on the political and foreign implications of international tourism, particularly from the American point of view.

Despite all our efforts to understand and systemize planning for recreation and tourism, a more philosophical role during planning and design is needed. The knowledge, experiences and methods captured in this text are useful not only technologically, but they contain the seeds .of deeper philosophical understanding. The reader is encouraged to keep an eye open for these deeper meanings as a way to advance understanding of these fields which contribute so much to the betterment of the human condition.

REFERENCES

Bakker, R.J. and A.E. Heil, 1988. De minicamping als nevenbedrijf voor boeren in Oost- en Midden-Brabant. Dr.-thesis Department of Land and Water Use, Agri-cultural University, Wageningen, The Netherlands.

Dam, D.J.M. van, 1988. Publiek Private Samenwerking en Landinrichtingsprojek-ten. Dr.-thesis Department of Land and Water Use, Agricultural University, Wageningen, The Netherlands.

Dietvorst, A.G.J., 1990. Changing planning strategies for recreation and tourism: Dutch experiences.

Edwards, A.R., 1983. Privatisering, een publieke keuze, V.N.G.-studies nr. 4, The Hague, The Netherlands.

Gelder, W.T. van, 1987. Bestuurlijke aspecten van en ervaringen met commerciali-sering op dagrecreatiecomplexen. Recr. en toerisme, 7/8.

Grontmij, 1987. Kurenpolder: onderzoek naar mogelijkheden voor commercialise-ring en privatisering, Concept, Breda, The Netherlands.

Hofhuis, P.J.M. Privatisering en toepassing profijtbeginsel. Uit: Open Universiteit. Economie van de collectieve sector 1, cursusdeel 3, blok 3, de overheids-uitgaven.

Jacsman, J., R.Ch. Schilter and W.A. Schmid, 1990. New Developments in Tourism and Recreation Planning in Switzerland. Inst. for Local, and Reg. and Nat. Planning ETH (Fed. Inst. of Technology) Zürich.

Kobes, N., 1988. Concurreren of subsidiëren. Een onderzoek naar de effecten van privatisering in de toeristisch-recreatieve sector, in het bijzonder in de Mookerplas en WRC Eysden. Dr.-thesis, Department of Land and Water Use, Agricultural University, Wageningen, The Netherlands.

Koopmans, L. and A.H.E.M. Wellink, 1987. Overheidsfinanciën, hoofdstuk 10: belastingen, zesde geheel herziene druk. Stenfert Kroese, Leiden, The Netherlands.

Kroese, E.P. and L.H.G. Slangen, 1987. Economische aspecten van de openluchtrecreatie, hoofdstuk 6: profijtbeginsel en inkomensverdeling. Department of Agr. Economics, Agricultural University, Wageningen, The Netherlands.

Lier, H.N. van, 1989. Wat wordt bedoeld met 'economisering'. Lezing PHLO-cursus 'Recreatievoorzieningen', Agricultural University, Wageningen, The Netherlands.

Mittmann, H.J., 1990. Recreation management within the multiple use concept of the United State Forest Service.

Philipsen, J.F.P., 1988. Begripsafbakening Economisering, Commercialisering, Privatisering en profijtbeginsel. Stencil, Agricultural University, Wageningen, The Netherlands.

Philipsen, J.F.P. and J.G Bakker, 1988. Privatisering van water- en hengelsport accommodaties. Een onderzoek naar processen en effecten. Concept Eindverslag, Werkgroep Recreatie, Agricultural University, Wageningen, The Netherlands.

Porter, J.A., 1990. Managing the planning and design process in Resort Communi-ties.

Tarlet, J., 1990. The French experiment -today's problematics- and new prospects.

Taylor, P.D., 1990. Nothing new, just ignorance revisited.

CHAPTER 3

Trends in world-wide tourism

by
Donald F. Holecek, Ph.D.,
Director, Michigan Travel, Tourism & Recreation Resource Center
Michigan State University
East Lansing, Michigan, U.S.A.

3.1   Introduction

David L. Edgell (1990) with the U. S. Travel and Tourism Administration, inspired in part by the words of Pope John Paul II, paints the following optimistic picture for tourism in the twenty-first century:

> *International tourism in the twenty-first century will be a major vehicle for fulfilling the aspirations of mankind in its quest for a higher quality of life a part of which will be "... facilitating more authentic social relationships between individuals..." and hopefully laying the groundwork for a peaceful society through global touristic contact. Tourism also has the potential to be one of the most important stimulants for global improvement in the social, cultural, economic, political and economic dimensions of future lifestyles. Finally, tourism will be a principal factor for creating greater understanding and goodwill and a primary ingredient for peace on earth. In my view, the highest purpose of tourism policy is to integrate the economic, political, cultural, intellectual and environmental benefits of tourism cohesively with people, destinations and countries in order to improve the global quality of life and provide a foundation for peace and prosperity.*

Edgell appears to see tourism as a form of panacea for all the world's ills, but in this paper, we will examine world economic trends for insights into the validity of Edgell's forecast of increased world prosperity through tourism industry development. The discussion will be organized under the following general headings: 1) trends in tourism consumption, 2) trends in political environments, 3) demand side trends, 4) supply side trends, and 5) projections of future consumption.

It is necessary to begin with a few brief definitions. Tourism is a term familiar to all, but one which almost defies simple definition. To facilitate analysis, economists often elect to define tourism to fit the data series available concerning tourist type activity. Unfortunately, available data series seldom provide enough detail to sort out tourism from closely related activities. Hence, the analyst is forced to modify the definition of tourism to match whatever data that might be available. Most commonly, tourism is lumped with travel in such data series. In a recent publication, Wharton Econometrics Forecasting Associates (1990) employed the following description of the travel and tourism industry:

PASSENGER TRANSPORTATION, land transportation by rail and highway; other land transport (e.g. sightseeing buses, taxicabs, rental cars); land transport support services, (toll roads, bridges, parking lots); water transportation, including ferries; air transport; and services incidental to transport (travel agencies, ship and aircraft brokers).
HOTELS, MOTELS AND OTHER LODGING PLACES, including rooming houses and camps.
RESTAURANTS, CAFES AND OTHER EATING AND DRINKING PLACES, including caterers as well as dining services on trains and other passenger transport facilities.
AND CULTURAL SERVICES, including theaters, museums, racing activities, amusement parks, golf courses, commercial sports, spas, radio and television broadcasting, motion picture production and distribution, theatrical production, libraries, botanical and zoological gardens.

The Wharton group's definition includes most readily identifiable types of businesses which service the needs of the traveling public. It does not include a category to capture traveler expenditures at retail establishments (for example, clothing stores).

Before proceeding with additional definitions, it is necessary to alert the reader, especially anyone less familiar with the limitations of tourism statistics, to the importance played by alternative definitions and measurements in interpreting trends in tourism expenditures. First, it is obvious that expenditure estimates derived by using definitions which do not include the same categories of expenditures will not be directly comparable. Estimates which include a retail shopping category, for example, are not directly comparable to the Wharton group's estimates since it did not account for tourists' retail shopping expenditures. Second, and less apparent, most analysts employ some formula for determining what share of total sales registered within a business type are attributable to tourists. Recall that the "passenger transportation" sector employed by the Wharton group included both "taxicabs" and "air transport". The latter is used almost exclusively for distant intercity purposes, thus all air transport sales could reasonably be credited to travel and tourism. Taxicabs, on the other hand, serve both the local population's need for local transport as well as the needs of tourists. Total taxicab sales, thus, can not be entirely credited to travel and tourism without exaggerating total travel and tourism expenditures. When viewed on a global scale, apportioning sales between tourists and others represents an immense challenge which is impossible to accomplish with complete objectivity given the limited availability of supporting data series. Despite comparability and subjectivity limitations, the data which are available are adequate for many purposes including that of providing the foundation for this chapter.

Economists define demand as the relationship between the quantity of a good demanded, in this case tourism, over a range of prices. This relationship is impacted (i.e. demand shifts) by factors such as income, education, tastes, age of consumer, and the like. Supply is defined as the relationship between the quantity of a good supplied over a range of prices. Supply shifts occur as a result of many forces including, for example, changes in production technologies. In the marketplace, demand and supply interact and result in some volume of a good being consumed. While supply and demand themselves are rarely measured or observed, consumption, the interplay of the two, can be measured. For example, many organizations report tourist arrivals and money spent by tourists. Such consumption data are generally available for many years, permitting identification of year to year and multiple year trends. These data alone are not sufficient to ascertain whether observed changes result from demand or supply shifts or from shifts in both supply and demand. While trends in consumption alone are useful in projecting the future course tourism may take, it is necessary to isolate and measure both underlying demand and supply shifting forces to enhance the reliability of such projections.

The economics of the world do not act in isolation of their political environments. Economics are grounded in basic principles of human behavior, but the rules guiding exchange are often set by individual countries or through international agreements. The future of international tourism is highly dependent upon trends in the political environments in which tourism must function. Hence, it is important to focus on these as well as trends in demand and supply as we attempt to project the future for international tourism.

3.2   Consumption trends

Possibly the best source of insight into the future is the past. Some statistical highlights for international tourism for the year 1989 are presented in Figure 3.1. Clearly, international tourism involves mass movements of people (400 million arrivals) and money ($200 billion U.S.). The flow of tourism receipts across international boundaries accounts for 7% of the world's exports. When coupled with domestic tourism, which dwarfs the international flow of travelers, total receipts captured by the tourism industry amounted to $2 trillion (U.S.) in 1989 (10% of world GNP), and the industry provides one in every sixteen jobs available in the worldwide economy. Finally, governments received $166 billion (U.S.) in taxes generated by tourism activity. Tourism is a strong force in the world economy and is beginning to command increasing attention.

```
*     Arrival: 400 Million People
*     Receipts (excl. fares): $ 200 Billion
*     Number of employees: 100 Million
*     1 in 16 Jobs Worldwide is Tourism Related
*     International and Domestic Receipts: $2 Trillion
*     Receipts represent 10% of World GNP
*     Tourism accounts for 7% of World Exports
*     Tourism generates $ 166 Billion in Tax Revenues
```

Figure 3.1   HIGHLIGHTS OF INTERNATIONAL TOURISM IN 1989

Source: Edgell, World Tourism Organization, Whatron, Econometric Forecasting Associate

The above statistics represent a snapshot of international tourism but provide no insight into recent and past tourism consumption trends. One indicator of tourism activity, international tourist arrivals, is depicted in Table 3.1 for the period 1950-1989. Growth has been drastic; nearly tripling during the 1950's, doubling during the 1960's, almost doubling again during the 1970's, and has grown by more than 40% during the 1980's. The interaction of demand and supply forces has resulted in strong growth in consumption of tourism throughout most of the post World War II period. However, the double digit growth experienced in the 1950-1970 period slowed over the subsequent two decades.

Fueled in part by inflation, international tourist receipts have grown even more quickly than have numbers of tourists. As can be seen in Table 3.2, tourist receipts more than tripled between 1950 and 1960, almost tripled again in the next decade, increased by almost a factor of six between 1970-80, and more than doubled in the 1980's. Between 1950 and 1989, international tourist receipts grew from $2.1 billion (U.S.) to $209.2 billion, an amazing hundred fold increase. During this same span of time, tourist arrivals increased by a factor of 16 indicating that per capital tourist spending increased by a factor of about 6.2.

The aggregate nature of existing data series, often leads to questions about what demand and supply forces are responsible for the rapid growth in consumption of tourism. In particular, many would attribute much of the growth to business related travel, yet as Figure 3.2 indicates only $637 billion (U.S.) or about one-third of the nearly $2 trillion in international plus domestic travel and tourism sales were derived from business and government sources. Personal travel generated more than twice the sales volume as did business/government related travel.

TABLE 3.1   INTERNATIONAL TOURIST ARRIVALS WORLDWIDE
(1950-1989).

Years	International tourism arrivals (thousands)	Percentage rate of change	Index (1950=100)
1950	25282.00	100.00	
1960	69296.00	174.09	274.09
1965	112729.00	51.32	445.89
1970	159690.00	36.36	631.64
1975	214357.00	30.47	847.86
1980	284841.00	29.36	1122.66
1981	288848.00	1.41	1142.50
1982	286780.00	-0.72	1134.32
1983	284173.00	-0.91	1124.01
1984	312434.00	9.94	1235.80
1985	326501.00	4.50	1291.44
1986	334543.00	2.46	1323.25
1987	361165.00	7.96	1428.55
1988	393160.00	8.86	1555.10
1989	405306.00	3.09	1603.14

Source: World Travel Organization

TABLE 3.2   INTERNATIONAL TOURIST RECEIPTS WORLDWIDE
(1950-1989).

Years	International tourism receipts (Million US$)	Percentage rate of change	Index (1950=100)
1950	2100	100.00	
1960	6867	227.00	327.00
1965	11604	55.54	552.57
1970	17900	45.64	852.38
1975	40702	89.23	1938.19
1980	102372	102.01	4874.86
1981	104309	1.89	4967.10
1982	98634	-5.44	4696.86
1983	48395	-0.24	4685.48
1984	109832	11.62	5230.10
1985	115027	4.73	5477.48
1986	138705	20.58	6605.00
1987	169539	22.23	8073.29
1988	194171	14.53	9246.24
1989	209155	7.72	9959.76

Source: World Tourism Organization

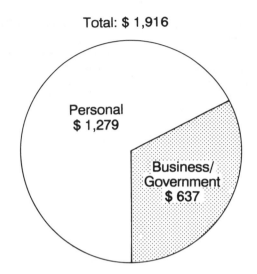

Total: $ 1,916

Personal
$ 1,279

Business/
Government
$ 637

Figure 3.2   DISTRIBUTION OF TOTAL WORLDWIDE 1987 TRAVEL
AND TOURISM SALES BETWEEN PERSONAL AND
BUSINESS/RELATED TRAVEL ($ Billion U.S.)

Source: Wharton Econometric Forecasting Associates

To complete this perspective on tourism consumption, it is useful to illustrate a few differences across countries and larger geographical regions. Figure 3.3 illustrates how residents of some of the world's most developed countries distribute household budgets between travel and other major categories of consumption. Travel accounts for between 12 and 19% of household expenditures in these countries. These figures illustrate the importance consumers in these countries place on travel and the vast potential for travel industry growth that would accompany economic growth in less developed countries of the world once their consumers incomes are adequate to meet basic needs for housing and food. The latter point is further illustrated by the information presented in Figure 3.4. One region, Europe, the most developed of the world's four geographic regions, accounts for almost 50% of total personal spending on travel and tourism. The least developed of the four regions, Africa and the Middle East, accounts for a mere 2% of worldwide spending on travel and tourism.

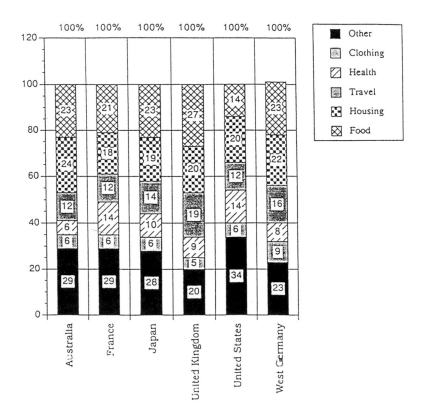

* Percentages in this column total 101%. Error introduced in source document possible due to rounding

Figure 3.3   DISTRIBUTION OF AVERAGE 1987 HOUSEHOLD SPENDING IN SELECTED COUNTRIES BY MAJOR EXPENDITURE CATEGORY

Source: Wharton Econometric Forecasting Associates

Population differences across these four regions, of course, account for a portion of the spending differences measured in these 1987 data, but they are not so great as to distract from the obvious implication these data suggest; economic growth in the least developed regions of the world is likely to spawn enormous increases in spending on travel and tourism both within and between regions.

Growth in tourism over the last decade has not been the same across the major regions of the world. Worldwide growth in international tourist arrivals between 1980-1989 ranged from a low of about 3% per year in the Middle East to a high of

24

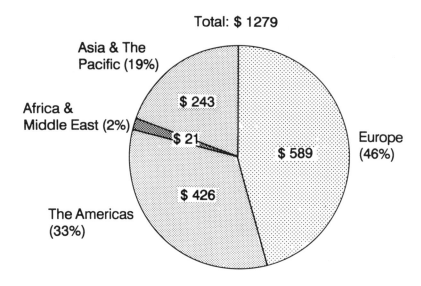

Total: $ 1279

Asia & The
Pacific (19%)

Africa &
Middle East (2%)

$ 243

$ 21

Europe
(46%)

$ 589

$ 426

The Americas
(33%)

Figure 3.4   REGIONAL DISTRIBUTION OF TOTAL 1987 PERSONAL
TRAVEL & TOURISM SPENDING ($ Billion U.S.)

Source: Wharton Econometric Forecasting Associates

about 9% per year in the East Asia/Pacific region. Worldwide growth in tourist
arrivals was about 4%. Change in arrivals over this period for the world, Americas,
and Europe are presented in Figure 3.5. With the exception of the period 1982-83,
the world and the dominant travel markets of Europe and the Americas have
experienced steady growth in arrivals. By 1989, Europe was registering 257 million
arrivals per year and the Americas accounted for nearly 80 million. International
tourist arrivals registered in the remaining four regions of the world are presented
in Figure 3.6. Note the vertical scales in these two figures have been modified for
illustrative purposes. Tourist arrivals in 1989 reached 44.5 million in East
Asia/Pacific, 13.3 million in Africa, 7.5 million in the Middle East, and 4.2 million
in South Asia. Over the period, the most rapid annual growth rate in the world,
over 9%, occurred in the East Asia/Pacific region fueled in part by rapid economic
growth in countries like Japan, Taiwan, and South Korea. Africa too experienced an
above average growth rate, over 7% during this period while the growth rate in the
Middle East, about 3%, lagged significantly behind the rest of the world partially
due to continuing military tension and threats of terrorism.

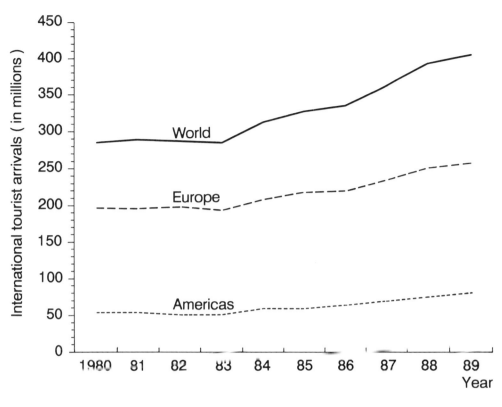

Figure 3.5   TRENDS IN INTERNATIONAL TOURIST ARRIVALS BY REGION, FOR THE PERIOD 1980-1989: WORLD, AMERICAS AND EUROPE

Source: World Tourism Organization

One final set of statistics will be presented to conclude this section. The relative importance of tourism in international trade is summarized in Figure 3.7. Tourism accounted for 7% of the world's export receipts in 1988. Tourism is most important to the export trade in Oceania, accounting for 11.6% of its export receipts. Above average tourism export receipts were collected in Africa (9.6%) and the Americas (8.9%). Europe's 6.9% lagged the world average (7%) slightly, and Asia's 4.9% placed it in last place. Asia's performance appears related more to above average development of its other industries (for example manufacturing) rather than to the performance of its tourism industry.

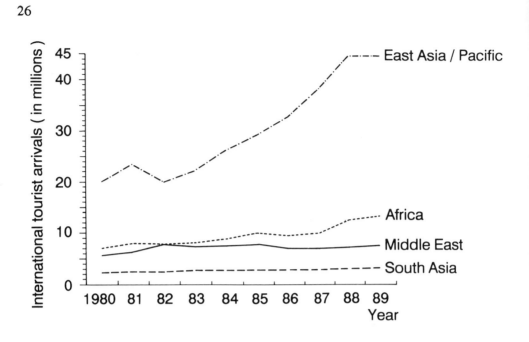

Figure 3.6   TRENDS IN INTERNATIONAL TOURIST ARRIVALS BY REGION FOR THE PERIOD 1980-1989: EAST ASIA/PACIFIC, SOUTH ASIA, MIDDLE EAST, AND AFRICA

Source: World Tourism Organization

3.3   Trends in political environments

The statistics provided in the above section clearly demonstrate that tourism has been growing over the last four decades and has increased in importance to the economies of the world and its major regions. Whether or not these trends continue and the relative speed at which changes will occur will depend on trends in a number of correlated areas. Science's ability to quantify and project trends in these areas is very limited and explicit models for relating them to changes in tourism have not as yet been developed. Thus, the approach that we will take at this juncture must be rather qualitative and subjective yielding incomplete and inconclusive results.

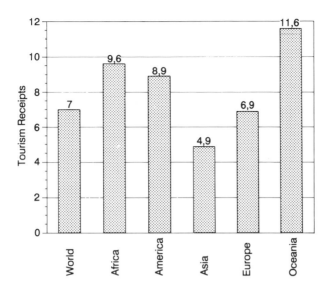

Figure 3.7   INTERNATIONAL TOURIST RECEIPTS IN 1988 AS A
PERCENTAGE OF TOTAL EXPORT RECEIPTS BY REGION

Source: World Tourism Organization

The political environment of the future will have a profound impact upon
worldwide tourism. Recent events can be used to illustrate three of the most
significant political environment trends which should command constant monitoring.
These are: the opening of Eastern Europe, the Middle East situation, and shifting
perceptions of tourism as an instrument of socio-economic change.

IRAQ INVASION OF KUWAIT - This event once again brings our attention to the
significant role this region of the world plays in international tourism. Obviously
this region's surplus of exportable oil is a major concern. The availability of
affordable fuel is crucial to tourism as well as to a wide range of other industries.
Political tensions in the Middle East which curtail exports of its oil reserves will
result in fuel costs which will in turn curtail travel volume directly by raising its
price and indirectly by reducing incomes of millions of potential travelers. Modest
increases in fuel prices spread over several years are unlikely to hamper growth of
tourism since many believe tourism demand is relatively inelastic or insensitive to
price changes. However, even temporary interruptions of supply can produce an
exaggerated response from consumers as has been demonstrated in previous energy
crises. A majority of travelers are unwilling to assume the risk, even if slight, of
not being able to obtain fuel required to complete a trip.

In addition to spawning abnormal fluctuations in energy costs and supply flows, events in the Middle East pose another threat to tourism which could have major and long term consequences. The increased threat of terrorism attacks on innocent travelers which Middle East tensions have fueled in the past will most likely yield a similar consumer response in the future. The majority of travelers simply won't make nonessential trips if they perceive terrorism as a significant probability. The perception of a terrorism threat often remains for years after the real risks have been eliminated. A region's touristic image may take years of effort and a great deal of expense to change especially if it has been the subject of dramatic negative news headlines over an extended period of time.

EASTERN EUROPE - The growth in international tourism is even more impressive when one remembers that billions of people still live in countries where freedom to travel is severely restricted. The opening of Eastern Europe to both inbound and outbound travel, if maintained, has the potential to profoundly impact international travel in the intermediate and longer term. If the trend toward removing travel restrictions extends to other countries, the potential for international tourism growth will be even greater.

The opening of Eastern Europe poses both challenges and opportunities for tourism industry growth. Similar conditions will be encountered in most countries of the world where travel has been restricted for decades. First, there is high pent-up demand for both inbound and outbound travel. Second, the ability for outbound travel is limited by low median resident income. Third, the capability for inbound travel is limited by availability of tourist facilities and the capital required to develop them and the country's general infrastructure, including a trained tourism industry labor force. Since modest investments in tourism related facilities can yield a quick return in terms of both international exchange credits and enhanced prospects for attracting foreign capital (e.g. tourists are also potential investors), many economic development specialists are recommending tourism as a target industry around which to build stagnate economies in these countries. Given the level of pent-up demand which exists and the prospects for both internal and external investment of new capital in these country's tourism industries and the availability of a large pool of trainable labor, the growth in freedom to travel should serve as a major stimulus to international travel later in this decade and beyond.

CHANGING PERCEPTIONS OF TOURISM - There is considerable evidence that the image of tourism as an industry is changing. Economists appear to be more appreciative of its role in economic development as suggested in the following quote:

The word is out, the economic growth industry of the 1990's is tourism, and every nation on earth wants a piece of the action. To the credit of the world's travel industry, both the private and public sectors have discovered the benefits that tourism can produce for a nation's economic and political standing. To the industry's detriment, however, there are few in either the public or private sector who appreciate the complexity of successful tourism development, and even fewer who are willing to invest the time or money to ensure that the benefits flowing from successful tourism development are maximized (Peterson, 1990). The wider acceptance of tourism as an economic development tool is related to growing acceptance of recreation and leisure as legitimate activities and to growing recognition of the importance of service industries in modern economies. As service industries have grown in importance as suppliers of jobs and as sources of export earnings, all service industries, including tourism, have gained in stature in economic circles.

Yet, the improvement of tourism's image extends far beyond economists. Socio-cultural specialists are also changing their perceptions of tourism. Tourism is seen as a factor in promoting world peace as suggested by Pope John Paul and a host of others. Their thesis is that the more countries exchange ideas, to include exchanging tourists, the more likely they are to understand each other and the less likely they are to want to wage war. Though less dramatic than reducing the prospects for war, tourism is viewed by many as an alternative to other forms of development which pose a greater threat to a country's cultural and natural resources. For example, the prospects of attracting tourist expenditures can result in the restoration of historically significant buildings and the protection of endangered species. As Peterson (1990) suggests, the improvements in tourism's image needs to be balanced by recognition of negative consequences of tourism development and investments to minimize them. Overall, trends in perceptions of the tourism industry bode well for its continuing future growth.

3.4   Demand side trends

When assessing prospects for an industry, economists often organize their analysis into trends in demand side and supply side factors which have impacted and are likely to continue to impact that industry. Herein, five demand side factors will be examined to arrive at an overall perspective of the role demand is likely to have on the future of international tourism.

INCOME - In order to afford to travel, consumers must earn adequate incomes. Income is linked to overall level of economic activity. The world's economy has experienced strong growth during most of the 1980's, but the economies of many countries of the world appear to be entering a period of slower growth. Thus, over the short term economic growth and incomes to fuel tourism industry growth don't appear promising. However, the world economy is likely to average continuing moderate growth over the longer term. This optimistic projection for the long term results from the prospects of a "peace dividend" resulting from declines in world tension, continuing benefits from technological advancement in many fields, and growth in free trade policies across the globe.

LEISURE TIME - Increases in leisure time available to consumers has been found to be strongly correlated to recreation participation to include pleasure travel. In general, leisure time available increases as economies expand and become more efficient. Continuing technological progress is providing expanding opportunities for capturing production efficiencies especially in the world's agriculture and manufacturing industries. The net result is that the future offers more leisure time for more people. Given the propensity of the consumer to allocate significant amounts of leisure time to international travel, the trend toward increases in leisure will support growth in both domestic and international tourism.

TRANSPORTATION - Demand for tourism is also correlated to access to comfortable and safe modes of transport. Cost is a factor in access to transport, and recent rapid increases in fuel prices will likely serve to reduce access and depress travel volume. In the longer term, technological developments are likely to improve fuel efficiencies in transport systems and yield alternative sources of energy to fuel the world's transportation systems. Technological developments are likely also which will increase the comfort and safety of transportation systems. The picture for the future of transportation is clouded by the growing discomfort associated with increased congestion of our transportation systems. The congestion problem will dampen the demand for international travel unless technological advances and improved planning are effective in offsetting congestion exacerbated by growing numbers of travelers.

COMMUNICATIONS - Advances in communications are not often considered in recreation/tourism demand studies, but we believe communications systems have a significant impact on demand for travel. The first and most obvious role of communications in travel demand is that of generating consumer awareness of and interest in new travel destinations. Private and public sector investment in tourism promotion in all forms of communications media is growing in recognition of the role communications plays in generating travel activity. Communications also plays a somewhat less recognized role in furthering demand for travel. It can be effectively employed to reduce the risk of having an undesirable experience. Access to effective communication systems provides the traveler with, for example, reliable information on facilities and services at potential destinations. Efficient and inexpensive communications also permits the traveler to remain in contact with family, doctors and business associates. This allows many people to travel who otherwise could not because they cannot be out of contact with what may be happening at home or in the office. Continuing rapid advances in communications technologies and their worldwide distribution will be a strong positive in the future demand for international travel.

FREEDOM - As already noted, the opening of borders to both inbound and outbound travel will be positive for international tourism if this tendency persists. Few would have predicted that the Berlin Wall would have crumbled in our lifetime, but it did in a few months after a few small cracks were identified by only a handful of observers. Such events are impossible to project into the very distant future, but for now at least the prospects for increasing freedom to travel are expanding.

In aggregate, the forces driving demand for tourism appear quite positive. Rising incomes and leisure time, improvements in transportation and communications systems, and growing freedom to travel suggest that consumers will be increasingly able to pursue their travel interests from almost anywhere in the world to almost anywhere else in the world.

3.5   Supply side trends

Economic theory holds that consumer demand for products and services is insatiable. Expressed differently, the world lacks the capital, labor and land to produce all the guns and butter that are demanded. The high debt accumulated in the U.S. and the limited stock of consumer products in the Soviet Union amply illustrate this economic principle. Continued growth in international tourism consumption will hinge not only on demand side trends but also on the world's ability and willingness to allocate the resources necessary to match growth in demand.

LAND - In the aggregate, the world has ample land resources (including land, water, forests, and air) to accommodate tourism industry growth, especially since this industry produces relatively low impacts on natural resources. However, as one begins to focus on specific tourism destinations, land resource constraints on growth become more evident. Development and population are often concentrated on a small land mass; oceans, lakes and rivers are often polluted; and, the world's natural areas are being rapidly converted for agriculture and other uses. The growing stress on the world's land has resulted in concern evident in resurgence of what we might label as the environmental movement. Airport expansions, golf course developments, and resort developments have received strong opposition from environmentalists in many regions of the world because they, for example, perceive natural resources as being scarce. Opposition to tourism development also often comes from other industry groups who see tourism as competition for land and from residents of communities whose citizens do not want to share their roads, restaurants and parks with tourists. The competition for land and related resources is not likely to abate in the future. For the tourism industry to garner the share of land required to grow, it will need to be sensitive to competing interests in its planning and development, and it will have to become more effective in the public relations and political arenas.

LABOR - Despite relatively high levels of unemployment in much of the U.S., one visits few cities where "now hiring" signs aren't displayed prominently in restaurants and hotels. Growth in service industries, including tourism, is exceeding the rate of growth in that component of the U.S. population, the young, which has traditionally filled entry level service industry jobs. Marriott Corporation's Bill Marriott and others have identified labor shortages as the major tourism industry challenge of this decade. Labor shortages are not, however, a universal problem. Some countries have a surplus of labor of the types most needed by the tourism industry. In many less developed countries, the labor challenge faced is at the management rather than entry levels. The nature of a tourism business requires a relatively high ratio of management to labor since many daily management decisions are best made at the lowest possible level within the corporate structure. Furthermore, a high percentage of a tourism business's employees come in direct contact with their customers, tourists, and must have required language and hospitality skills to effectively represent their employers. Until recently, the world's educational systems were not widely involved in preparing students for careers in the tourism industry. Recently, there has been rapid development in tourism programs within higher education in the U.S. and other countries. Major corporations and even smaller tourism businesses have become aggressive in educating service employees and in developing incentive programs to retain employees at all levels.

CAPITAL - Without growing access to capital, the tourism industry will not be able to grow fast enough to keep pace with growing demand. As noted earlier, the improved image of tourism as an attractive economic development vehicle will help in the quest for external investment capital. Yet, in many regions of the world, limited public infrastructure such as roads, telephone systems, and water/sanitary systems will severely constrain the development of a sizeable tourism industry. In such settings, tourism development is still possible via targeting the most adventuresome traveler or through developing more self-contained tourism complexes. The latter require relatively large amounts of investment capital and are most often the subjects of local resident resentment because the benefits from such development are more likely to accrue to outside investors while the negative impacts are borne by the indigenous population.

While political environments and demand side trends appear to be quite positive for tourism industry growth, supply side trends in the aggregate are less positive. Environmental concerns, labor availability, and access to capital appear likely to hinder growth in some regions. Supply will likely be slow to respond to demand shifts leading to frequent local shortages and surpluses. Even a sluggish response on the supply side is unlikely to keep tourism from growing at a strong pace over the next few decades.

3.6 Projections for the future

The World Tourism Organization (WTO) has prepared forecasts for international tourism to the year 2000. Its forecast for international tourist arrivals and receipts is presented in Table 3. Projections from both the 1950-89 and the 1980-89 trends are included. By the year 2000, arrivals worldwide are projected to increase by over 110% from the number registered in 1989 using the 1950-89 trend and by nearly 60% using the 1980-89 trend. Receipts by the year 2000 are projected to increase by about 300% from the amount registered in 1989 using the 1950-89 trend and by about 150% using the 1980-89 trend. The WTO forecast also includes projections of relative market share for four world regions: Africa, Americas, Europe and Asia/Oceania. The greatest increase in market share is projected for the Asia/Oceania region while Europe is projected to experience the greatest loss in market share.

The combined quantitative and qualitative projections of trends presented herein suggest continued growth in international tourism at least through the year 2000. These projections do not provide for the possibility of either major shifts in political circumstances such as a major war or for a natural disaster with worldwide impact. Excluding such unthinkable events, it is, thus, our feeling that the tourism market will continue to grow. This does not mean that tourism will grow everywhere. Local success will vary widely depending on such factors as: marketing effectiveness, quality of planning, and conditions confronted in local political environments.

34

TABLE 3.3   WORLD   TOURISM   ORGANIZATION'S,   INTERNATIONAL
TOURISM FORECAST BY REGION, 1989-2000.

	Forecast			
	Based on growth rate 1950-1989		Based on growth rate 1980-1989	
	1995	2000	1995	2000
WORLD Arrivals (mn) Receipts ($bn)	641 444	956 845	515 353	637 527
AFRICA Arrivals (mn) Receipts ($bn)	26 13	40 22	23 10	32 14
AMERICAS Arrivals (mn) Receipts ($bn)	114 103	154 172	103 95	128 146
EUROPE Arrivals (mn) Receipts ($bn)	376 219	530 403	294 152	338 206
ASIA/OCEANIA Arrivals (mn) Receipts ($bn)	125 109	232 247	95 86	140 161

REFERENCES

Edgell SR., D.L., 1990. Charting a Course for International Tourism in the Nineties: An Agenda for Managers and Executives. U.S. Dept. of Commerce. U.S. Travel and Tourism Administration and Economic Development Administration, Washington, D.C.

Peterson, E., 1990. Tourism: Planning, Promotion and Marketing. The Courier No. 122. Brussels, Belgium.

Wharton Econometric Forecasting Associates, 1990. The Contribution of the World Travel and Tourism Industry to the Global Economy. American Express Travel Related Services Company, Inc., New York, N.Y.

CHAPTER 4

French planning for tourism and recreation - new perspectives

by
Jean Tarlet
Doctor in geography and planning at the Service for Watermanagement
and Land Use Planning in the Provence Region,
Aix-en-Provence, France

The evolution of the tourist trade and the leisure industry in France is most significant in so far as the country has one of the world's oldest tourist industries. Today the industry is confronted with prospects of important developments with the opening of the great European market in 1993.

In fact, tourism in France began during the second half of the nineteenth century. It flourished in favorable conditions and grew considerably. This growth is linked with the history of the country and of Europe. Since the end of the Second World War, the tourist and leisure industry has played such an important part in the French economy (France is the fourth largest tourist country in the world with thirty-five million visitors a year; that is to say, ten per cent of world tourism) that it is even said to be a leisure civilization. The idea of leisure activities goes beyond the idea of tourism since it includes the short week-end as well as daily leisure activities.

Tourism has indeed created landscapes. It also has destroyed others, and for about fifteen years tourist planning in general has been criticized. The quantitative and qualitative change in demand in the population, has compelled authorities (state and local communities) as well as property developers to be conscious of the deficiencies in the system, and modify their ways of intervening. The prospect of the opening of frontiers for a single European market after 1992 also implies adjustments from all those responsible for tourism development. Today, the French tourist and leisure industry is therefore in evolution. Based on new methods and new concepts, interesting information is being derived for those responsible for planning and landscape architecture not only in France, but in other countries as well.

4.1   Geographical characteristics--great diversity, a dense human occupation

France is not a large country compared to some. It has a surface area of 551.000 square kilometers (Corsica included). That is to say France is fourteen times smaller

36

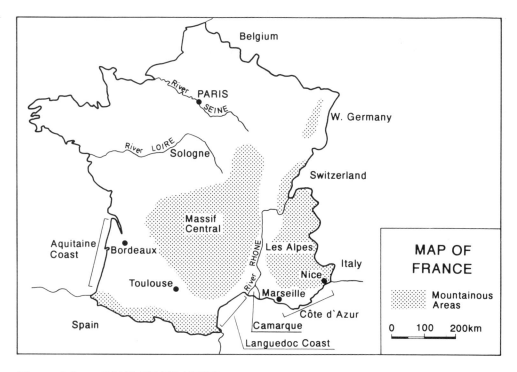

Figure 4.1   MAP OF FRANCE

than the United States (seventeen times if Alaska is included). But, the effect of the landscape and the climate make it a very diversified country. It is relatively mountainous - in the center, the Massif Central (see figure 4.1) separates the vast plains of the North and the West from the Rhone Valley and the Languedoc in the South. The Alps witch bound it on the East culminating at 4807 m (Mont Blanc) provide an important skiing area. The Pyrénées, forming the border with Spain, are much more arid. These reliefs bring about the partitioning of the country.

The coastlines also are varied. France is the sole European country with three coastal fronts. Its 3.200 kilometers of coastline include cliffs in the Côte d'Azur and Normandy, sandy beaches in the Languedoc and the Aquitaine, and jagged, rocky coasts in Britanny, marshlands in the Rhône Delta.

The climate variations also add to the country's diversity. The Mediterranean coast with its mild climate all year long, and its hot and dry summers has often been compared with the Californian climate. The West coasts have a cooler climate, mild and damp, and becoming fresher north of the Loire, which forms a real climatic border. The North East of the country shows greater temperature ranges.

Winters can be quite cold and snowy, especially in the mountain regions. The Northern half of the country is to a considerable extent more rainy.

The human occupation of the country is dense with 57 million inhabitants (that is to say on average almost one hundred inhabitants to a square kilometer, excluding 20% of the land which is forests and mountains). All the plains have been cultivated and developed, for centuries. In the favorable areas numerous towns with rich histories have built up progressively. Among these are Paris (almost ten million inhabitants), along with Lyon and Marseille (almost a million inhabitants), and other "regional capitals" with over 400.000 inhabitants, such as Lille, Nantes, Toulouse, Bordeaux and Nice.

It is important to point out that this complex combination of ancient human settlements in original, characteristic sites, has given rise to a range of prestigious tourist sites:

- Le Mont St Michel, a medieval, pyramid shaped town crowned with a monastery on a rocky islet isolated in the middle of a bay;
- Islands like Porquerolles or Ouessant;
- The marshlands reserves for horses, bulls and birds such as the Camargue, or the Sologne ponds, silent and misty, ideal for shooting and fishing;
- The castles of La Loire dominating the vast plain between the river and the forest;
- The vast harbors of Brest, Arcachon, Toulon;
- Lastly, all manner of ancient fortresses and villages, like les Baux perched on hill tops, or overlooking old towns.

In short, France is a diversified country with a wealth of tourist advantages, including:

- an exceptionally diversified coast-line;
- mountain ranges figuring among the highest in Europe;
- an exceptional patrimony of sites and monuments;
- Paris;
- its location as a crossroads between Northern and Southern Europe, between the Nordic, Atlantic and Mediterranean countries.

On the other hand, there is a dense human occupation, and living space is relatively scarce. The famous Côte d'Azur, for example, can expand only along a narrow coastal plain between the sea and the foothills of the Southern Alps.

4.2   One of the oldest tourist traditions

It is from 1860 onwards that an "aristocratic" tourism was organized on the Côte d'Azur (Cannes, Nice, Menton) thanks to the railway. This tourism is principally an English phenomenon and gives its name to the splendid boulevard in Nice, "la promenade des Anglais".

Early in the twentieth century the fashion for travelling and holidays became more and more popular. It was the phenomenon of a rich bourgeoisie which settled, for example, on the Normandy coast (Deauville) or in Aquitaine (Biarritz). Students toured by bicycle. They camped, and bathing began on the beaches.

The year 1936 was a key date in the history of French tourism. The new socialist government, among numerous social laws, made it compulsory to grant two weeks' paid holidays to all wage earners, leading to a real tourist explosion.

The Second World War halted this process, of course, but it picked up again little by little when peace was re-established. The old rural society, war-weary in 1945, was to enjoy thirty years of prosperity which the French called "les trente glorieuses", a period when a prodigious economic and landscape transformation was to take place. For tourism the sixties represent a period of exceptional expansion, which can be explained by a whole series of factors (1 to 4):

1) Until then France was a country of farmers. During the 1960's it became a country of city-dwellers (chart enclosed). This rural exodus brought about an important steady decline in the rural population, as table 4.1 shows.

TABLE 4.1   SHIFT OF THE FRENCH POPULATION FROM RURAL TO URBAN AREAS IN PERCENTAGES (FROM 1801-1982)*

	1801	1901	1946	1975	1982
population in urban areas	23 %	41 %	53 %	68 %	73 %
population in rural areas	77 %	59 %	47 %	32 %	27 %

*   adapted from Roger Sue "Vivre en l'an 2000" ed. Albin Michel - Paris 1985 - quoted in "Temps et Loisirs" - A.P.A.S. oeuvres sociales - January, 1986.

Modern towns where people were forced to live to find work, and new conditions of industrial work in particular provoked among the city dwellers a violent desire to flee from urban areas. They could:

- be in contact with nature again, and to feel a freer, more genuine life;
- flee from pollution and lack of comfort (noise, second rate accommodations, stress) in town life. The themes of fresh air, pure water and unspoiled landscapes became major ones in commercials; and
- rediscover their rural or family roots with a certain nostalgia, hence week-end or holiday visits to grandparents who stayed in the village.

2) The extremely rapid rise in the standard of living enabled families to give an increasing part of their budget to tourism. In the same way the development of the car industry allowed an easier "way of escape" from daily surroundings.

3) The extension of paid holidays constitutes an essential factor to the French attitude about tourism. From two weeks a year in 1936 it went to three, then to four weeks in 1969. In 1981 the socialist government, in office again, granted a fifth week's holiday, and lowered the retirement age from sixty-five to sixty, and in certain cases, temporarily, to fifty-seven.

4) Lastly, there arose among authorities a new consciousness of the tourist phenomenon and its social and economic impact. While limited, this consciousness entails state support for tourism development. It is shown in particular by means of planning. Planning in France plays only an indicative role. That is, it only emphasizes the economic priorities. Essentially it enables investments to be directed. However it also expresses the goodwill of the state in the economic field.

As Pasqualini and Jacquot (1989) explain, the French state has practically intervened since the end of the Second World War. The first national plan (1948-52) made provisions for the extension of transport infrastructures, and the creation and modernization of hotel facilities. The next plans dealt with the development of camping, social tourism for the underprivileged, and tourism in the rural milieu. The 5th plan (1966-1970) marked a new direction. It stressed:

. the expansion of winter sports, giving rise to the creation of winter resorts;
. the development of sailing;
. the development, under the impulse of the state, of the Languedoc-Roussillon coast-line; and
. the creation of the first natural parks, both national and regional.

Lastly, the 6th plan (1971-1975) dealt with the development of the Aquitaine coast and for the first time showed some concern for environmental protection and the fight against pollution. In fact, the rapid development of tourist activities in the absence of protective measures had already brought about a certain number of negative consequences of which the French were only just beginning to be aware.

4.3   Recreation product development and the difficulties they impose--an assessment

The French effort in tourist and leisure development continued, therefore, for almost a century, and more particularly during the last thirty years, according to a rising demand linked with the evolution of the French economy and society. The country's most interesting sites are now fully developed, so it is possible to analyze past actions and draw conclusions about the future from France's development efforts.

4.3.1   Coastline development

The seashore was the first destination for the touring public. By the 1910s and 1920s the beaches were regularly invaded by holiday-goers. But if the first generation on the Côte d'Azur were winter tourists, fleeing from the London fog, by the early 20th century, the trend had reversed. Today, it is the summer period that constitutes the tourist season.

At first, swimming and beach games constituted the primary recreation attitudes at the seashore. Then, in the 1960s, rapid advances in innovative recreation technology broadened the array of recreation opportunities. Affordable and portable boats, underwater diving equipment, water skiing equipment, and growth in ownership of pleasure craft, caused spectacular changes in tourism markets.

In addition, wind surfing became a high-use activity, because it required little space and is inexpensive. Today, the number of pleasure boats is estimated at 620,000 (of all sizes - motor and sail) to which 440,000 sail-boards must be added (1985 figures). To satisfy the seasonal migrations of millions of people today (in 1987, for example, 25 million tourist stays to the French coast were estimated), an important development of infrastructure became necessary:

- Transport infrastructures such as the Nice airport which have had to be extended repeatedly with land reclaimed from the sea on the Côte d'Azur. Various facilities such as car parks also have been set up on reclaimed land on the sea coast.

- The number of ports reserved for sailing has been increased more and more artificially, and sometimes without sufficient preliminary studies leading to the facilities' being choked with sand.

- Numerous activities connected with sailing, have been set up (including repairs, shops, berths, and security services).

- Around this, there is the local community to consider, with the many social and economic impacts it now faces.

Accommodations in the coastal zone have been considerably developed and diversified. Over the years the grand hotels of the beginning of the century (of which the Negresco in Nice is still the epitome) were turned into flats, or underwent other changes. Their rich clientele have abandoned them for other sites. On their place a more modest hotel business has been extensively developed (especially two-star hotels), as well as camping and caravaning sites offering a wide range of prices according to the facilities available.

Artificial villages have been sometimes created around sailing ports, forming small tourist resorts. Cazes et al. (1990) give two examples: The "Marina baie des Anges" near Cannes with 1300 flats, accommodating about 4,000 people. The facilities' three huge tower blocks, 70 meters high and along the shore, pretentiously dwarf the surrounding landscape.

Another example is Port Deauville, which Cazes describes as "a port with 1,250 moorage rings, 312 "marinas", 450 flats, hotels, restaurants, bars, shops...". In so doing, Cazes speaks of an explosion in shore use. According to him, the communes on the French coastline cram 14 per cent of the population onto less than 4 per cent of the territory. The density of this development reaches 250 inhabitants per square kilometer, and has absorbed 20 per cent of the population increase since 1936. On the Côte d'Azur 90 per cent of the coastline has been urbanized, with over 40 per cent elsewhere. 2,800 kilometers of the coast line have thus been built up out of a total of 3,200 kilometers.

Obviously, such upheavals have brought about important consequences concerning the landscape and the environment. If the positive aspects are undeniable (the creation of industrial activities and jobs, and the chance of more pleasant holidays) the negative aspects must also be pointed out:

- the pressure on the environment, and
- the seasonal nature of these activities.

42

The pressure on the environment

The coastlines are, by definition, a limited, fragile zone. According to St. Marc (1974), most tourist resorts are included in a 3 kilometers wide strip next to the sea, as well as the greater part of the infrastructure, especially harbor infrastructure. Closer to the shore, on a second 100 meters wide strip, on both sides of the shore line, one finds most of the holiday makers and their amusements. This 200 meters wide strip, all in all, "is the ideal" zone for bathing and games: it is also the place where the land and sea interreact.

Could this very narrow coast strip, in a country where living space is scarce, happily accept such urbanization? Perhaps if sufficient, preliminary studies had been made, and if a sufficiently strict, and properly respected legislation had found a way of protecting the environment. This has not been the case.

- Starting from the first urban centers (fishing villages or market towns) urbanization stretched out in a line along the "strategic" area--that is to say the contact line between land and sea. Then, later, it gained ground inland. Going from the sea towards the land frequently the following zoning is found:

. a seascape covering a few hundred meters where bathers, sail boards, pleasure boats and various water sports are active,

. a part of this coast reclaimed for the building of infrastructure, sometimes only to extend the beach which can't expand inland. A third of the underwater coastline is said to have been made artificially on the Côte d'Azur,

. the beaches themselves, sometimes private ones which are overcrowded, especially in summer. The Côte d'Azur has at present reached its saturation point,

. directly behind the beach, luxury hotels and catering establishments with their own car parks,

. further back from this sea-front more modest hotel accommodations (the lower the prices, the farther away) and the town itself with its tertiary industries, businesses and services.

In the final zone, on the outskirts of the town, sites with a sea-view have been colonized by family dwellings, which are often very costly. The lowest regions, formerly farm-lands, are equipped very modestly with camping and caravaning sites. The criterion of organization is principally the distance from the sea.

Going from the land towards the sea, in the opposite direction, a certain number of negative factors are found in conjunction with this sort of urbanization:

1) The progressive disappearance of some of the best farm-land which is often alluvial ground (in particular the deltas of small coastal rivers). Due to the presence of the sea the micro climate is very favorable. There is no frost and it is at times very sunny (Côte d'Azur). There are therefore often excellent farm-lands which disappear, and this conflict between agriculture and tourism is one of the principal criticisms of this sort of urbanization. In other sectors there are sites of great ecological value (rivers, marshlands) which are purely and simply destroyed. The excessive trampling of the dunes in the North of France has even at times brought about the disappearance of the vegetation fixing the sand then, in consequence, the beginning of the burying of coastal villages has taken place, as the dunes start moving again.

2) A second consequence was the building of a "concrete wall" along the beaches creating a barrier between the rural zones and the sea, broken only where the relief does not permit building. French beaches, with their variety, have witnessed the construction, from the beach of a sort of straight wall formed by the hotel buildings. These structures are supposed to be "at the water's edge" and, in the end, have aided the destruction of the original landscape in which the charm of the site lays.

3) The architectural quality of this new urbanization should also be questioned. The construction of blocks of multi-story flats by powerful groups of private contractors, and built as rapidly as possible, reveals a mediocre aesthetic quality and is ill-adapted to the sites. Apart from a few rare exceptions this type of architecture is generally considered inappropriate.

4) The saturation of beaches at peak periods, from July 10th to August 10th approximately, is less and less accepted by the general public.

5) The elimination of waste, especially liquid waste, also creates a problem. It can only be discharged in the sea while the "purification" plants don't always work satisfactorily. In fact the plants, which are always difficult to control correctly, have difficulty in absorbing the variations in population, and purification is not always thorough. Hyères, for example, changes from 40,000 inhabitants to 250,000 in the summer. In spite of the construction of longer sewage outlet pipes, the cleanliness of the beaches is hard to maintain and some of them had to be forbidden to holiday-makers in the past. However, important improvements have been made during the last decade in this field, and a warning system works pretty well, at present, on French beaches.

6) The coastline is more and more saturated with users, especially on the few hundred meters on the coast itself. It isn't so much the bathers as the different water sports, sea scooters for example. Serious accidents can occur during the holidays, such as those caused by rapid launches colliding with bathers lying on floating mattresses.

7) Also the impact of water sports, the concentration of which is increasing incessantly, on the underwater environment, should be underlined.

It is well known that the coast ecosystem is based on the beds of sea plants, zostera or eelgrass beds in the Atlantic, and posidonia beds in the Mediterranean. These plant beds are today, principally on the Mediterranean coast, subject to multiple aggressions linked up with leisure activities. The impact on them includes:

- The fill-ins on the sea-board which have covered over a part of them,

- The dumping of large quantities of liquid waste and other toxins during the summer which increases the turbidity of the sea-water. Photosynthesis can no longer take place and the beds wither.

- Boats anchors scraping the sea-bed which also attack plant beds.

- Lastly, the increase in underwater fishermen which has led to a real scarcity of fish, especially in the mediterranean which is poor in fish populations already.

The seasonal aspect
The second category of negative points concerns the seasonal nature of coastal tourism. The coastline suffers an extreme concentration of tourists for a limited period of two months. At the school holiday period all means of transport including planes, trains and motorways are saturated. During the two crucial months, and above all the peak period from July 10th to August 10th, car traffic becomes extremely difficult in the coast sectors. Hotel keepers have only a few weeks to make their establishment pay, and so do most of the services and businesses. A great number of second homes are open only for four weeks of the year and closed for the other eleven months. Most of the sailing boats are out only a few days a year, and they tie-up room in the ports the rest of the time.

Such a state of affairs constitutes an economic aberration. Generally speaking, the chaotic urbanization and, more generally still, the planning (or non-planning) of the French coast zone has become one of the principal problems of national development in France. It is stating the evidence to say that from the 1970s onwards state intervention was needed and the coastal spaces should have been organized by radical means.

The main state-planned projects
    By the 1960s the state had become aware of the economic benefits of tourism and the need to organize its expansion. Two great projects were then undertaken: the planning of the Languedoc-Roussillon and Aquitaine coastlines.

Languedoc-Roussillon
    The planning of the Languedoc-Roussillon coast was put into the hands of an "Interdepartmental Mission" created in 1963, but it was to associate local communities, and private promoters. It affected a coastline of 180 km, 120 km of which was a low sandy coast, without trees, with numerous ponds, formed by small, coastal rivers and separated from the sea by an off-shore bar of sand dunes. The proliferation of mosquitoes made this region, which was very hot and dry in the summer, very unpleasant indeed. At the same time the road network was very poor.

    The Mission aimed at displaying this region to its best advantage while respecting the environment, so as to develop the local economy and create jobs. The principal idea was to build "integrated" tourist resorts which were big enough to constitute new small towns (each with from 100 to 120,000 inhabitants) but separated by stretches of coastline which had been left intact. Thus, there would be no road along the coastline, and travel would be assured by an inland motorway, with spur roads leading to the coast.

    The program, which has become definite by 1964, consisted of the following (see figure 4.2 and figure 4.3):

* The definition of an intervention perimeter along the coastline, 10 to 20 km deep. It affected 4 departments and 67 rural districts.
* The realization of the development plans on three levels:
    - the plan for regional planning,
    - the local plans ("communes"),
    - the mass-plans for the new resorts.
* The acquisition of 4,500 hectares (11,250 acres).
* The establishment of all infrastructure.
* The total clearing of mosquitoes from the coast by chemical treatments, in particular, and by removing the marshlands.
* The reafforestation of 6,000 hectares (15,000 acres).
* The adaptation of inland lakes for water sports.
* The creation of 20 sailing harbors each with a capacity of 400 to 2,000 boats.

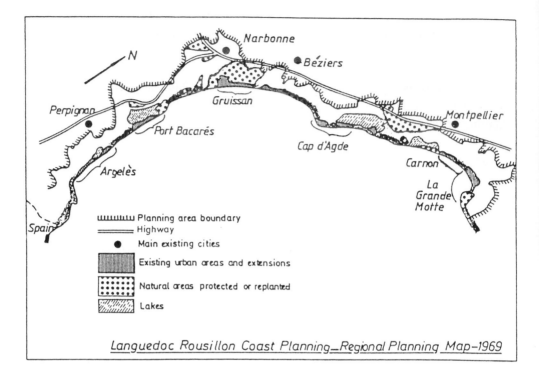

Figure 4.2   THE REGIONAL PLAN OF "LE LANGUEDOC ROUSSILLON"

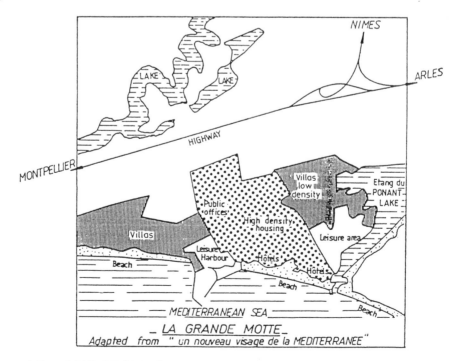

Figure 4.3   MAP OF "LA GRANDE MOTTE"

* The creation of six new tourist resorts, veritable towns, linked as far as possible with existing towns such as "La Grande Motte", "Cap d'Agde", "Gruissan", and "Barcarès-Leucate".

These new resorts, planned for all sorts of clientele, from the most modest to the richest, were to contain 250,000 beds. It was also planned to add 150,000 beds in old towns, that is to say a total of 400,000 beds, accommodating 2 millions tourists a year.

The Aquitaine coast

According to similar principles "the inter-regional Aquitaine coastline planning project", created in 1967, was to realize a smaller plan concerning a coastline of 200 km. This site was a long, straight, sandy beach in front of dunes reafforested during the last century, cutting off a chain of ponds. Here also, the aim was to promote the economy while respecting "the natural capital" and avoiding anarchic urbanization.

The program, involving 620,000 hectares (1,550,000 acres), 450,000 of which were forest lands with 114 rural communes, was supposed to be fully-comprehensive, planning the whole of the regional activities. With this aim in view, procedures of landownership control affecting 160,000 hectares (400,000 acres) have taken place. It is one of the most important operations of landownership control undertaken in France.

The essential principle was also the planning in depth with a spurroad off the motorway with the development of water tourism, focused on the 80,000 hectares of ponds and lakes (see figure 4.4).

Finally, separated by green belts (secteurs d'équilibre naturel) "areas of natural harmony", nine new resorts were to be realized: "le Verdon", "Hourtin", and "Lacanau", for example. Planned as dense urbanization, they had to be "grafted" on to the small towns already existing, like Arcachon. 220,000 beds were planned which could accommodate 800,000 tourists. In this way, it was hoped to diminish the share of the second homes, and the gnawing away of living space. The plan also aimed to limit camping to 26 per cent, to plan with the inhabitant's participation, and to extend the busy period to five or six months of the year instead of two.

First results

After more than two decades, the results seem to be positive, but somewhat irregular. The dangers for the environment have been limited. The architecture in the new resorts is not always unanimously approved. From an economic point of view the plan has been a success on the Languedoc-Roussillon littoral: three

48

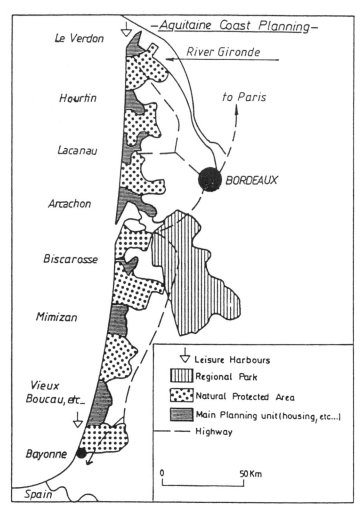

Figure 4.4   THE "AQUITAINE COAST" PLAN

millions tourists visit it each year. The development of the resorts on the Aquitaine coast has been, on the contrary, much slower. On the whole, this planning has succeeded but it only affects a coastline of 400 kms.

4.3.2   The development of the mountain zone

The history of, and the difficulties in developing the French mountain zone have been analyzed meticulously by Cazes et al. (1990). Today, France is the world leader in "winter tourism" with about 450 winter resorts and 1,500 square km of well equipped, skiable ground out of 1,900 available. Nevertheless, a series of episodes marked the development of an environment very different from the littoral - difficult, and at times hostile, yet, all the while, fragile.

At the end of last century, the Alps underwent the progressive, uncontrolled development of summer tourism, resulting in the high class hotels being closed in the winter. Sports activities and games were numerous and included facilities for skating, gambling, and skiing. All in all summer developments of a hundred years ago required little space, and made few encroachments on the fragile mountain zone.

The development of winter sports resorts took place in the 20th century and approximately three generations of winter resorts can be distinguished:

- The 1st generation showed little organized development (the Val d'Isère, equipped with a plan for urbanization in 1935, is an exception). This development concerned winter tourism, almost exclusively, and they were built by private companies, in particular the railways.

- The 2nd generation appeared after 1945 and was better planned. It was the local communities, and the department councils rather than the district councils which took development into hand, assuring landownership control by preparing a development scheme themselves. Almost as a harbinger to today the idea arose to equip glaciers for summer skiing.

- Lastly the winter sports boom in the 1960s, more recent but perhaps more rapid than coastline development, brought about the construction of a 3rd generation of winter resorts. These were considered by Cazes et al. (1990) as "a last hope for an economic take-off" in the High Mountain sectors, abandoned by the traditional activities. These winter resorts, of which Avoriaz, les Menuires and Pra-Loup are examples, are the largest of all. They are at the same time "integrated" resorts resulting from a development doctrine carefully worked out. Today, they are now concerned solely with sports skiing according to public demand. State intervention is more noticeable, allowing, in particular, the opening up of transport facilities, but one single developer coordinates the whole project.

Some logical principles for the mountain development were defined at this point; for example they demand a fairly strict zoning:
   - ski runs all located on the northern side of the slopes;
   - construction is on flat and if possible sunny ground, extending vertically, with concentrated urban style housing conditions;
   - there is separation of the vehicular traffic from pedestrian ways; and
   - car parks are located on the outskirts of the development.

A development doctrine was thus slowly defined, and it had necessarily taken into account the mistakes made in the past.

Mixed results

The tourist development of the mountain zone, such as it has been carried out in France for the last thirty years, shows mixed results. Among the positive aspects, to be sure, is the revitalization of the High Mountain zone, which, otherwise, would have been doomed to depopulation. In addition, important road infrastructure has opened up the high valleys. As a result, the mountain zone now benefits from new sources of income related directly or indirectly to sports. It is estimated that 300,000 jobs have been created, with 25 per cent being permanent. Many seasonal jobs help local farmers, who, otherwise, couldn't earn a living without mixing their professional activities.

But important negative aspects are also to be noted. The preliminary studies were at times insufficient, in projecting snow coverage, for example, and many winter resorts in the Southern Alps have experienced serious difficulties due to the lack of snow in the last few years. In the same way, market studies were at times insufficient, resulting in a series of difficulties, and environmental analyses were inadequate. The result has included attacks on the environmentally sensitive mountains. At times the landscape has been completely changed needlessly. Elsewhere, the construction of artificial winter resorts was set up with an imposing, ill-conceived architecture, offering to the visitors a mineral scenery instead of small, more intimate chalets. Lastly, they did not manage to avoid, especially in the oldest resorts, an anarchic, unorganized development of constructions with exaggerated use of living spaces on the community outskirts. The lack of integration with existing villages has also been criticized, and this lack of contact with the existing populations tends to emphasize the artificial aspect of certain resorts, both from a social as from an architectural point of view.

Also serious deficiencies in the economic program should be pointed out. This tourist expansion in the mountains did not help to reduce the decrease in farming. Seasonal jobs are still badly paid and the socioeconomic effects hoped for are less significant than was foreseen. At the present time, this market in France has reached the saturation point: The demand no longer increases; it is estimated that at least a quarter of the resorts have serious management difficulties; and, investors are not keen on investing in this kind of operation anymore.

4.3.3   Tourism in the rural environment

It is not easy to give an exact definition of the rural space in France. Theoretically, among the 36,000 French communes, the rural ones are those with less than 2000 inhabitants. But, they must be distinguished from the suburban sectors. Situated on the outskirts of an important town, these suburbs have totally different characteristics, including immense, latent possibilities for holiday visiting. The actual rural space represents, however, 9/10s of the country's territory, and includes the sectors of middle-sized mountains with their particular characteristics.

It is not a question of France's rural spaces being empty spaces, but of a very old inhabited, cultivated space concerned since the middle of the last century with the rural exodus from the countryside to the towns. Naturally, rural areas are a less attractive environment than the coastline, the mountains or even the towns. They offer fewer assets and up till now have been little exploited. They also are extremely diversified.

Yet, the French countryside, as does almost all of Europe, possesses an old civilization whose ways of adapting itself to the environment are varied and often picturesque, offering a great variety of architectural and cultural traditions. A number of picturesque sites have been and can be exploited, because they have a marked personality and have kept an "ancient" look. As examples, we cite:

- sites of geographical interest--valleys with gorges (Ardèche, Tarn, Lot),
- monuments (Pont du Gard, old castles, abbeys or small villages perched in the mountains),
- national or regional natural parks, created recently (Vanoise, Mercantour, Volcans d'Auvergne) (about 30 in France) or natural reserves (Camargue) (about 29).

Certain sectors lump several communes in an intermunicipal association with many different goals. This is done to share the exploitation of their various possibilities, such as rambles on foot or on horseback, shooting, fishing, fine restaurants, and the like.

Up till the 1950s, the rural exodus, often entailed brief returns, at holiday times, of families coming to see their grandparents or wanting to keep in contact with their native villages. Corsica has a reputation for this sort of visit.

The development, by the authorities, of tourism and leisure activities in rural districts is recent and dates principally from the 1970s. There are several reasons behind it. First, the state wanted to put a stop to the rural exodus and the desertification of the land. The national development policy for the land was based, at the beginning, on the willingness to give a harmonious development to the French regions. Tourism, during the last decades, seemed to be the best means of reaching this objective.

Second, the saturation of the coastline persuaded holidaymakers to look for other sectors, such as small lakes. Lastly, holidaying in rural districts is much less expensive and therefore better for families of modest income, provided that there are leisure activities to offer.

Thus the development of the rural zones was undertaken, specially by the communes, on a modest scale in accordance with their limited means. It was a question of constructing swimming pools, sports grounds (small playing fields, bowling grounds), a village hall (for shows, and a few film shows), signposting and the upkeep of the tourist areas such as river banks with fishing grounds.

At the same time accommodation facilities were to be developed. The old country hotel accommodation was modernized, with light facilities added, such as camping and caravaning sites. New accommodation possibilities have been created these last twenty years, such as "rural gîtes" (rooms to-let in farmhouses), boarding houses, bed and breakfast accommodations, and buffet suppers for residents, "tables d'hôtes".

Likewise, organizations like "family holiday villages", "V.V.F.", have set up small hotel complexes with various services available (group activities, child care centers) in rural sectors especially for less wealthy families. 60,000 beds in 130 villages were set up by this organization alone.

Modest but encouraging results
This recent development of tourism and leisure activities in the rural environment has not excluded errors and difficulties.

First of all, it is not possible everywhere, because there exists a "downward spiral" in the very underpopulated zones. When the demographic decline is too great, there is no longer a sufficient number of dynamic personalities, or enough local willingness to take charge of local development. At times parochialism prevents the communes or the neighboring valleys from working together due to excessive mutual mistrust. Now it appears that the policy of evenly distributing funds was a mistake. Cazes et al. (1990) affirm that a minimum threshold of concentration and complementarity of holiday facilities is necessary for the real success of a tourist village. We can take as an example the village of Moustiers in the Southern Alps which possessed several assets:

- beautiful surroundings,
- an old renowned pottery craft,
- several lakes.

In many cases, the refusal of the villages to band together in a common organization was the principal cause of the failure. Others, who managed to work together to create country resorts and obtain credits from the authorities succeeded in checking the decline of their communes.

A second difficulty arose from the uncontrolled increase of second homes, estimated at over 11 million beds in France. This phenomenon is in keeping with the French character insofar as they are very attached to private property. This sort of "gnawing away" of the countryside led to considerable monopolization of living space and degradation of the landscape, characterized by the unwarranted privatization of certain village paths, the setting up of fences, and the increase in fire risks in the Mediterranean zone. It entailed a supplementary, financial weight coming from the extension of the public service networks (water, energy, telephone), which it demanded. Finally, this spreading has become the source of numerous conflicts with other activities (agriculture, sports and entertainment activities) concerning land use.

Conflicts sometimes break out in the rural environment from the clash between town and country people. When holiday centers or important accommodation facilities are being built, many people believe that the number of holidaymakers should not outnumber that of the permanent residents if frictions are to be avoided. Certain people even think that the proportion of holidaymakers should not be over 40 %. But this proportion is difficult to respect in villages of no more than 100 or 200 inhabitants.

To sum up, in spite of the difficulties, the policy of developing rural tourism, essentially devised to revitalize declining regions, has had successes of limited importance. This is true particularly in comparison to the enormous tourist influx into the coast regions or the mountains.

4.3.4   *Tourism and leisure activities in the urban and suburban environment*

It is in this field that leisure activities are different from tourism. In fact these activities can be practiced daily or on week-ends, whereas tourism implies an absence of a least a few days. Different investigations have revealed the practices of French people. Since 1945 a great effort has been made in developing urban facilities including stadia, swimming pools, tennis clubs and various sporting centers.

However, there exists a specifically urban tourism centering around visits to famous monuments. But, it is largely a question of foreigners. On holiday periods French towns lose most of their inhabitants who go to the beaches in the summer, and the mountains in the winter. But many foreigners take advantage of this exodus, to visit French towns, and Paris in particular. Paris in the month of August has the reputation of being a town emptied of its inhabitants, but almost all the languages in the world can be heard there. Among the 13 most visited monuments and museums in France, 11 of them are to be found in Paris: the Pompidou center, and the Eiffel Tower, the Notre-Dame cathedral, the Louvre Museum, and Versailles, among them. The other two monuments are le Mont St Michel and Chambord castle. This sort of tourism is linked with the policy of the protection and restoration of cultural heritage.

Water-cure tourism and the casinos associated with it, should also be mentioned. Until the early 1980s urban and suburban tourism consisted of these fairly traditional forms of activities. In the following years, however, this sector is likely to be developed like the others.

4.4 Today's problems--new perspectives

Tourist activity, as we have seen, has become more and more important since the beginning of the century. It poses a certain number of problems which have not yet been perfectly solved. What is more, it seems certain that the volume of this activity is going to take another step forward on a world scale as well as a strictly French one. These perspectives create a new situation with new, related difficulties.

4.4.1 The prospects of global development

Certain experts foresee a doubling of world tourism in the next 15 years. As regards the Mediterranean region alone, which is already first in world tourism, the number of tourists could pass from 100 millions today, to 400 or 700 million in 2025. A part of this increase concerns the French Mediterranean front.

In addition, the creation of the single European market, is also likely to increase tourist activity in Europe and in France, in particular. Europe has a potential 330 millions inhabitants, 50 per cent of whom go on holiday now. Some people reckon that this figure will be 60 per cent at the end of the century.

We can also expect, to some extent, the continued shortening of the work week,which is 39 hours in France. Finally, longer life expectancy and medical advances imply an increase in the demand coming from older people.

These perspectives, as a whole, imply the pursuit of the growth of the tourist activity. These expectations are in keeping with those of the United Nations according to which tourism, in the year 2000, will be the principal service industry and will have to respond to the increase in free time.

4.4.2   The new context in legal and administrative terms

For France, as for many other European countries, one of the most remarkable economic events of the last few years, was the unification of the E.E.C. countries. In order to define the tourist policy, and the sharing out of tourist funding, the European authorities are going to have more and more influence. Inevitably this implies a change in the parts played by the various administrative levels in France.

Only sufficiently large regional bodies can really hold a dialogue with the Brussels authorities*. French regions, whose administrative power is of recent date, have come out stronger. The communes or the association of communes remain the basic unit. On the other hand, the small departments and to a certain extent the French state tend to lose a part of their economic power to the advantage of the E.C. on the one hand, and the regions on the other.

In addition, a totally new legislation was set up recently. The 1982-83 decentralization laws reinforced the development powers, given to the local authorities, that is to say the communes and the regions.

Since 1984, planning has no longer been defined only on a national level. The facilities and development programs are the subject of five year plans negotiated between the state and the regions, with the financial share of each member specified yearly. The different operations, especially in the tourist field, tend to be the object of a complex organization in which at times each of the five administrative levels (Europe included) intervenes, along with private developers.

Lastly, the many difficulties posed by development and by tourism in particular, brought about the setting up by the state of more protective legislation. It is a question now of "the coastline law" and "the mountains law" completed by "hill development plans" and "plans for sea development" (S.M.V.M.). Very few of these have been drawn up, however.

The 1986 law for the protection and development of the coastline aimed at protecting sites, and the ecological balance by prohibiting roads and buildings on the coastline, by obliging to build in continuity with the existing towns, and by establishing a list of spaces to be protected, such as dunes, islands, and marshlands.

---

* Continental France is divided in 22 regions, 95 "départements", and about 36,000 "communes" (= the basic administrative unit).

The 1985 law concerning the development and protection of the mountains was not confined to protective measures but also aimed at economic development by planning for:

- local authority initiative and control;
- the valorization of farming, forestry, craft industries, industry and energy;
- urbanization in continuity with existing villages;
- statutory definition of "multiactivity"; and
- the establishment of hillside committees.

4.4.3 The evolution in the clientele's demand

It was inevitable that the future clientele, in its formation as in its tastes and expectations, should be different from the old ones. Therefore, there must be taken into account not only the reasons for the dissatisfaction of the present clientele, but also the tastes of future clientele insofar as they can be foreseen.

1. The setting

One of the first pieces of information coming from enquiries made to the clientele, was the refusal to accept the gigantic proportions of certain coastal or mountain resorts. People do not go on holiday to find themselves with the same sort of housing conditions they have just left in town. As Trigano (1989), the founder of the "Mediterranean Club" stated, "overstressed town dwellers need good holiday breaks, they need to get back to friendly village life, where they have time, space, and can adopt a different lifestyle". Future tourists will probably prefer a setting which is smaller, more human, more intimate--even a family setting, in fact.

2. Holiday programs

Numerous surveys revealed that changes in holiday plans (the distance, the length, the frequency) bound up with the changes in modern life (more spare time, a drop in transport prices, the economic crisis, etc.) seemed to apply to the whole of Europe. To sum up, people are traveling shorter distances, for a shorter time, more often.

3. The activities

One of the principal expectations of tourists is their concern with activities practiced during the holidays. According to a saying which has become popular, the tourist is no longer happy sunbathing stupidly on a beach. This truism is the partial cause of tourist stagnation at the Côte d'Azur, where too much advantage was taken of its favored site with no effort to improve the organization of activities. According to Thurot "passive holidays" have given way to "active holidays", and this change in demand corresponds with a certain number of related factors:

Contact with nature
This refers to active contact, that is: from foot excursions and horse-riding to "eco-sports" such as cycling or jogging. The coming into fashion of golf seems to go with this trend.

New sports
Young people are attracted by the discovery of activities, sometimes completely unknown, such as mountain parachuting, hanggliding or ultra light planes. Canoeing and rafting are also in fashion.

Cultural expectations
Cultural discovery is in keeping with new expectations: the discovery of a country, a region, monuments, cultural traditions (local fêtes) or old customs (in particular craft industries), even local ecology (plants, fossils, wildlife). An interest is also taken in history, and in the research of one's own roots.

Health
"Fitness courses" (sea water therapy for example) are growing in popularity. They are intended in particular, for managerial staff suffering from stress.

We also find in the surveys certain expectations concerning games, communication with other people, with other cultures. People want cozy family holidays, which explains the French ideal of having a second home.

When sea is near at hand, water sports are still in greatest demand, but they are no longer enough. Today, a range of other activities is wished for in the same site.

In fact it can be noted that all these expectations are often linked. Holiday expectations appear to be the negative of daily life which seems artificial and compartmentalized.

The main conclusion is that today's tourists are harder to please especially regarding services (day-care centers for example) and cultural activities. They want holidays offering new discoveries: nature, a new sport, other people. All this, they want already organized and presented as consumer goods that they do not wish to prepare by themselves, because they do not have the time. They often prefer a fixed-price program which promises them a stay coming up to all these expectations. We are therefore witnessing an explosion of new holiday plans, as a means to the realization of these expectations.

4.5   The answers: a new approach--new "products"

Now, everybody is conscious of the economic importance of the tourist and leisure activity sector. But in order to continue taking advantage of this activity trend, it is indispensable to follow the changes in the general context as well as the demands of the clientele. Since the 1980s, the state has no longer launched great planned operations. It intervenes, however, as we have seen, by an effort of organization, protection, and encouragement through legislation and development schemes and with consultation with the regions.

Now, it is up to the promoters, private or public (companies controlled by the state through the medium of nationalized banks or public services like the S.N.C.F. (the French rail company) and local authorities to define and promote new approaches and new services to avoid the mistakes of the recent past, while at the same time delivering the expectations of the clientele.

4.5.1   The general organization

The dense concentration of tourism in space and in time presents one of the principal difficulties. A joint effort from private promoters and the state was thus made along the following lines: extending holidays to ensure a longer peak season, especially for winter tourism; avoiding overloading roads, (a source of accidents); and ensuring a better return from the facilities over a longer period.

The French territory was thus divided into 3 zones for which the dates of school holidays, especially in the winter, were different for each zone. Up till now this policy has not given totally satisfactory results. On the other hand, an attempt has been made to find out-of-season programs such as business courses, training courses, and periods reserved for the elderly.

It also has been necessary to develop areas under-exploited up till now, that is to say the rural areas, because the principal development of tourism has remained concentrated on the coastline, the Northern Alps and in Paris.

Time-sharing schemes have been developed at the seaside and in the mountains to limit individual constructions and lower costs. Thus one may be the owner of a week's holidays, for a life time, in one resort. In certain cases one is even the owner of a studio apartment. But one has neither the service charges nor the responsibility of administration.

Lastly, today electronic booking systems for hotel rooms are coming on-line, especially at the department scale.

4.5.2   More serious studies

Today we see the need to carry out a series of preliminary studies before any tourist operation is taken into account. These studies are more complex and more thorough than two decades ago, for example:

- Resource studies, on a departmental or regional scale, undertaken by regional or departmental commissions, and including a list of sites, monuments, rivers and lakes, cultural traditions and festivals likely to help tourist development.

- Biophysics environmental studies, the maximum capacity of an area (a very difficult notion to define), especially concerning mountains. Environmental impact assessments obligatory since 1977, have to be carried out in a serious way once the operations have been defined and the sites chosen.

- Studies of global physical and economic planning to be effected, if possible, at the very beginning of a development operation so as to ensure its integration in local and regional development without neglecting purely socioeconomic aspects such as jobs or the existing population. Methods called "ecological planning" can be a precious aid at this stage of studies.

Such studies were used, for instance, in the preparation of the principal scheme for the development and urbanization of the Toulon region and for the development of the artificial lake of Ste Croix*. The U.N.O. (its European Economic Commission) insists on the need for integrated planning from a physical, environmental, socio-economic and cultural point of view.

- Marketing and management analyses
As surprising as it may seem, this sort of study had often been neglected up till now. Yet it seems obvious that it is necessary to prepare any tourist development with an accurate analysis of the number of visitors planned for, maintenance costs, profitability, as well as a definition of ways of managing and selling the area. For public projects, direct administration by communal services, for example, turned out to be inadequate. The creation of a semi-public company ("société d'économie mixte - S.E.M.") proved to be preferable.

Recent failures of three "theme parks" brought to light certain inadequacies in this sort of study. The theme park "zygofolis" in Nice for instance went bankrupt soon after its creation. The Mirapolis park near Paris, has been less visited than

---

*   These studies had been carried out by the "Société du Canal de Provence et d'Aménagement de la région provençale".

previously planned, and consequently has had to reduce the number of its employees. The concern in following the changes in clientele and their expectations inspired the creation, on a regional or departmental level, of a certain number of organizations such as tourist observatories, carrying out annual surveys on tourist flow, expectations expressed, and expenses. The publication of these figures is a precious aid for public and private organizers.

4.5.3 New services--new programs grouped together

Totally new programs appeared recently and there is no end to innovation in this field. Besides traditional programs all sorts of new services are making their appearance. People now talk about health tourism (water cures, health clubs, sea-water therapy), cultural tourism (including economic or industrial discoveries), business congresses, gastronomic, and photographic tourism (safari photos in natural parks to study wildlife), shooting, or fishing tourism.

We must also mention river tourism (by boat on quiet canals), and horse and cart excursions on country roads. Chains of horse riding centers have been developed for horse-riding tourism. There is, however, a lack of trails for pedestrian or horse-riding tourism and certain associations have recently taken in hand the signposting of long distance walks, with appropriate maps. Other new sports also are appearing such as country motor cross, and mountain biking.

- Thinking about the products
But, above all, instead of thinking of hotel accommodations, or facilities, or games, there is a tendency to think of products, which are fully organized beforehand with a fixed price. The list includes:
- tours with different stopover points including visits to castles or fine restaurants. These may be organized as "a week of discovery", with everything under the same rubric (integration);
- fixed prices for "shooting-plane" (especially to foreign countries)*.

Lastly, new resorts have been created on a smaller scale, avoiding gigantic proportions and respecting the environment. Buildings of no more than 2 to 4 stories, surrounded by greenery, each one within an original style are the result. They offer, on site, sports and cultural activities, swimming pools, golf, tennis and a wide range of activities and sports. Providers know that tourists will not really practice the sports they demand, but they are happy just having the opportunity. Out of 100 people wanting horse-riding hardly a dozen will in fact practice it. This reality avoids the appearance of oversized facilities. In an atmosphere of complete

---

* Package tour prices for plane and hunting holidays.

freedom (no obligations), holiday makers can take a training course in pottery, weaving, painting or ceramics, or can go off in search of a fossil deposit or a natural reserve for studying wildlife.

Under this model services will be at the holiday makers' disposition. Such services can include appropriate catering facilities, day-care centers, cars for hire, and even audiovisual facilities for training courses. These are fully organized centers where accommodation, as Guiral (1988) wrote, is only part of a whole, this is a new leisure philosophy with integrated resorts or multi-leisure centers. The result is an association of accommodations and activities in an environment and an atmosphere made as attractive as possible. This is the simple idea which has made the fortune of the famous "Mediterranean Club" and of the association "village, vacances, familles" (V.V.F.), designed on a more limited scale, and for people of modest income.

4.5.4   New achievements adapted to every sort of environment

On the coastline
New trends towards small integrated unities have materialized in certain new achievements such as the lakeside city called "Port Grimaud", and designed by the architect F. Spoerry. He is not keen on modern architecture. As he explained in an interview with Martine Leventer ("Le Monde" - 17th November 1989) "the aim of architecture is to make man happy". So he tries to rediscover "the magic of old towns", by planning living spaces" which give the impression of having grown up from the long maturing of coincidences and time". Thus he adopts the old village style, traditional, architectural details to create lively, friendly resorts which do not oppress their inhabitants. He detests constructions which are inhuman, oversized, and aggressive. Port Grimaud is a real French village typical of the Mediterranean coast with real fishermen's cottages, of not more than two stories, compact and not lined up, each with its boat moored in front as in Venice. Similarly, in New York when building "Port Liberty", he drew his inspiration from New England architecture. In spite of criticism for "pastiche", his theories on soft architecture have already interested the Prince of Wales (see figure 4.5).

In the mountains
The village of Ceillac in the Queyras, a region of the Southern Alps, drew its inspiration, a long time ago, from principles that we are rediscovering today, thanks to its mayor Philippe Lamour. A "village resort" was developed with less pretentious facilities, integrated more successfully into the landscape. Above all, development was genuinely controlled by the local community, especially through

62

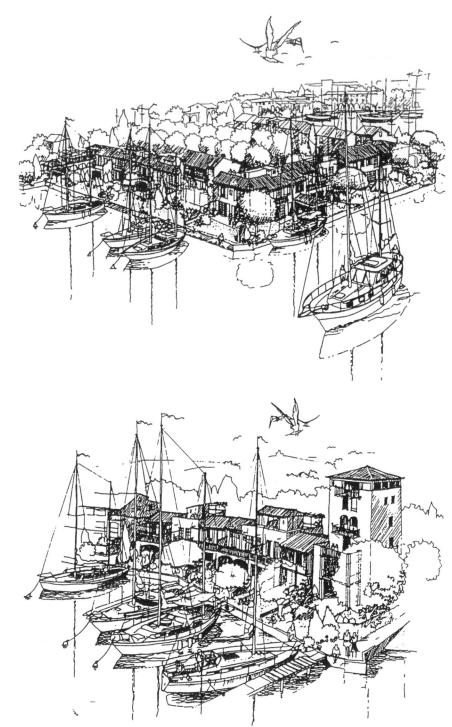

Figure 4.5   PORT GRIMAUD (TOP) AND PORT GRIMAUD III (BOTTOM)
(CITÉ LACUSTRE) AS DESIGNED BY F. SPOERRY (1988)

control of land ownership, the participation of inhabitants, and integration of the resort into local life. In the same way, facilities were designed to be less mono functional, and were able to work in the summertime as well.

In rural zones

The authorities have foreseen, that by taking into account the evolution in agricultural technology, nearly 600,000 or half of all French farmers, will have given up the profession by the year 2000. Such an outcome is likely to speed up the rural depopulation which seemed to have been stopped in recent years. It has become evident to authorities that tourism is the one activity likely to create jobs in the rural zone and put a stop to rural depopulation. One sort of development which is becoming more widespread is the small holiday center. Typically, these projects are centered around small lakes (from 4 hectares (10 acres)) with a range of accommodation facilities in the neighborhood (campsites, caravan sites, bungalows for hire, hotels, restaurants) and a varied range of services, activities, sports and games. It is thought that the success of such a development is first of all due to its being near water, but credit also is given to the mixed range of sports, games and services offered to tourists.

Besides wind surfing and boating when the dimensions of the lake permit, tourists are offered water games, horse riding, children's games, tennis, bowling, ping pong, archery for beginners, karate, mini-golf, and other recreation opportunities. However, the aid of state and local authorities is indispensable since such holiday centers are not very profitable, especially if the grounds have to be bought. Among the 150 holiday centers in rural zones the majority were financed by the whole of the local communities. Typically, there are complicated deals associating the commune or the communal association, the "département", the region, the state and even European funds, with each party contributing a certain amount. At times, the state intervenes through semi public companies, SOMIVAL in the Massif Central, the "Society of Canal de Provence and the development of the Provence region" (SCP) in the Southern Alps, is an example. Figure 4.6 illustrates a typical layout for this sort of center.

The difficulties in projecting success with this type of complex come from the market analyses. The relative uncertain number of visitors planned-for can entail oversized facilities making management even more difficult (the minimum cost of such a center is 30 million francs). We must also mention, in Southern France in particular, that technical problems such as eutrophication accelerate management costs.

The similarity of stereotyped facilities could end up turning away the clientele. But, it is to be hoped that in each case one will try to find and promote a

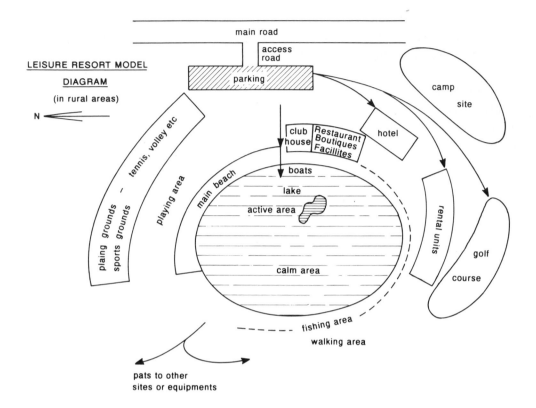

Figure 4.6   LEISURE RESORT MODEL DIAGRAM (IN RURAL AREAS)

completely original plan for each center, based on unusual sports such as canoeing, or the quality of its catering, or its golf course, or its health center, for example. Each resort must be individualized.

Suburban activities

The expansion of leisure activities has led to the development of a new sort of activity in suburban areas. Up till now, suburban sectors were equipped with playing fields, stadiums, tennis courts and public parks only. Now, largely artificial activities have been organized according to present demand. Near at-hand activities for the week-end or Wednesday afternoons, requiring the closeness of population reservoirs in the big towns or on the coastline is an important factor.

There are 3 principal aspects:
* Grounds equipped for new sports such as karting, moto cross, or golf.
* Theme Parks, inaugurated in the 1960s and exemplified by the Thoiry animal park and the OK Corral. The fashion is very recent, and the biggest one is Euro Disneyland in Marnes la Vallée, near Paris.
* Other diversified recreation centers, such as:
  - the Futuroscope (near Poitiers) (science, discoveries)
  - the smurfs' park (a fantasy world)
  - Asterix (pseudo-historical)
  - the Aqua-lands such as Marineland in Antibes
  - the Center Parks, equipped with bathing facilities in a tropical atmosphere under a glass dome, with housing facilities such as hotels or bungalows. These Center Parks are planned to function in the winter and receive guests for short holidays or on the week-end. Only one exists in France in Verneuil sur Avre, and built in 1988 (The "Aquaboulevard", in Paris, is similar). These latest centers demand large investments, but despite the costs they seem to have difficulty establishing themselves in France, unlike those in the U.S.A. or the Netherlands.

Only 40 theme parks are to be found in France. And, it should be noted that the development of such parks is tied to advances in transportation, giving rapid access from towns and cities.

Leisure activities in town
Some authorities are beginning to think that leisure activities should be brought nearer to the town. After all it is the most favorable site for daily leisure activities, confirmed by the U.N.O. report (1988) on planning for leisure activities and tourism. If urban degradation causes migration to the country, renewal should reduce this. Towns too, have their heritages which can be exploited. On recent years the restoration of town centers, and the exploiting of monuments and architectural heritage has been undertaken. Many festivals have been organized in urban areas such as Aix-en-Provence, Orange, and Avignon.

But new urbanism has to be practically defined. Reintroducing nature into the town is very expensive. Yet, it is not impossible when use is made of grounds belonging to disused factories or when private parks are opened to the public. The setting up of sports centers, children's playing grounds and green areas is then possible. Some people are in favor of green trails, allowing people to go to leisure activity sites by bicycle or on foot. Riverside areas sometimes are easily cleared to provide pleasant walks.

Renovation of poor town centers into lively pedestrian streets containing luxury shops, restaurants or displays has been undertaken in many French towns. Car-use has also been limited to the benefit of public transport, and carparks have been built on the outskirts and the town center declared a "pedestrian zone" (Bilanges, 1987). Even the architecture can be influenced by these new conceptions. An example is the construction of private housing estates with tiny gardens (from 100 to 200 m^2). These developments help to dissuade people from going away on week-ends without increasing excessively the need for urban living space.

4.6   Summary and conclusions

Far from being at the end of its evolution, the field of tourism and leisure foresees spectacular developments (see appendix 4.1). Fortunately, many lessons have been drawn from the mistakes of the last few decades.

Great efforts must still be made, however, regarding the organization of the profession, and public authority support. Gilbert Trigano, President of the Mediterranean Club, speaks of a "disorganized" profession which is not yet ready to think in terms of products.

Because demand changes quickly, one must understand the tastes of the clientele and if possible foresee them. Clientele are more demanding and will not accept the barrack - buildings solutions of the 1960s. These consumers are also very sensitive to crisis such as the Tchernobyl accident, and terrorist murder attacks which helped cause the re-establishment of visas to visit France, resulting in a considerable drop in tourism. Progress must be made in several fields:

- the staggering of holidays;
- the development of studies concerning resources, physical planning and ecological analyses;
- the use of market analyses, even if it is difficult to define precisely what socioeconomic repercussions are to be hoped for;
- management;
- the choice of new products and services, the quality of the facilities and the commercial organization; and,
- the coordination of research between European countries.

Above all, one must be prepared for constant changes and adaptations according to public demand.

It must also be recognized that in the present French context, the role of the state and local communities seems to be essential. Their role has become more important and will become more so, because the authorities will inevitably have the power to make more decisions as living space becomes increasingly scarce. On that subject, decentralization has not always had positive results.

Many sites have been built. Today the strategical stakes seem to be the exploitation of rural space, with few facilities. Rural sites, on the whole, are less attractive. Therefore, artificial holiday centers, attractive enough to draw the public, must be developed. This implies heavy investments and great efforts to attract a large enough clientele. Because of the uncertainties, private developers will not invest on their own. Thus, considerable aid, support and management are needed from the public authorities.

Tourism calls into question the whole policy of national planning and development. A supplementary effort of control is therefore necessary. Better quality urbanism would have discouraged these gigantic week-end leavings just to get away from ugly urban surroundings. The leisure civilization has contributed, perhaps more than elsewhere, to shaping, with new styles of living, the modern French landscape. In the future this evolution must be more harmonious.

REFERENCES

Aménagement Touristique du Littoral Languedoc-Roussillon, 1969. In: "Techniques et architecture" numéro spécial 2-31, Paris.

Barbier, B., 1987. "Problems of Tourism" (symposium) Tourism Institute, Warsaw (Poland).

Bilanges, J.P., 1987. "Géographie Touristique de la France" éd. B.P.I., Paris.

Cazes, G., R. Lanquar, and Y. Raynouard, 1990. "L'Aménagement Touristique", P.U.F. "Que sais-je", Paris, 3ème édition.

Dominati, J., P. Guiral, J.J. Descamps, Y. Raynouard, G. Trigano, 1988. "L'enjeu touristique", Economica, Paris.

"Espaces", 1985. Revue, numéro 72, Paris.

L'Aménagement de la côte aquitaine, 1971. La Documentation française, Paris n° 265-266.

Memento du Tourisme, 1989. Ministère du Tourisme. La Documentation française, Paris.

O.N.U., Commission Economique pour l'Europe, 1988. "La Planification de l'espace pour les Loisirs et le Tourisme dans les pays de la région de la C.E.E.". O.N.U., Genève.

Pasqualini, J.P., and B. Jacquot, 1989. "Tourismes", éd. Dunod-Bordas, Paris.

St. Marc, P., 1974. "Le devenir du littoral: un choix de société". In: "L'architecture d'aujourdhui" no. spécial "Habiter la mer", no. 175.

Trigano, G., 1989. "Forces et faiblesses du Tourisme français". In: "Problèmes économiques". La Documentation Française, n° 2108, Paris.

APPENDIX 4.1   FIGURES   ON   RECREATION   AND   TOURISM DEVELOPMENT IN FRANCE

Holiday departure rates [*]:   1964 - 43.6 per cent
                                    1987 - 58.2 per cent
For 1987 this percentage equals 32 million French people and 927 million of recreation days.

Periods [**]:   - 2/3 in July and August: in summer: 708,700,000 days,
                                                   in winter: 225,700,000 days.
The rest in September, the spring holidays, and at Christmas time.

Length of the stays [**]:
A fortnight on average, (twice a year), approximately 29 days a year in all.

Departure rate according to the categories [**]:
The lowest: -   farmers: 34 per cent,
                     -    retired people: 47 per cent.
The highest: -   senior managers and professional people: 88 per cent.
People go less often on holiday from 20 to 24 years old and over 50 years old.
According to the wage brackets the rate goes from 27 per cent to 91 per cent.

Accommodations [**]

Hotels	1 million	beds	Second homes	11,300,000 beds
Camping, caravans	2.4 million	beds	Holidays villages	250,000 beds
Gîtes and guest rooms	177,000	beds	Youth hostels	20,000 beds

Percentage of days spent on holiday in France according to the type of lodging (1984) [***]

At parents' and friends'	39.4 %	Second home	13.7 %
Tent or caravan	17.4 %	Hotel	5.9 %
Renting	16 %	Others	7.6 %

Destination [****]

The coastline	42 %	The town	7.9 %
The mountains	19.6 %	Tours	3.6 %
The rural sector	26.5 %		

Transport means [****]

	Winter 1983-84	Summer 1984
- by car	77 %	83.3 %
- by train	16.7 %	10.2 %
- by bus	2.6 %	2.7 %
- by plane	2.8 %	2.2 %
- by boat etc...	0.9 %	1.6 %

---

[*]   Trigano, 1989.
[**]   "Momento du Tourisme", 1989.
[***]   adapted from J.P. Bilanges - see references
[****] I.N.S.E.E., 1986.

CHAPTER 5

Planning for outdoor recreation and tourism in the Netherlands

by
Adri Dietvorst
Professor and Chairman Center for Recreation Studies
Dept. of Physical Planning and Rural Development, Agricultural University;
Head, Dept. of Outdoor Recreation and Tourism
The Winand Staring Centre for Integrated Land, Soil and Water Research;
Wageningen, The Netherlands

5.1   Introduction

The Dutch have a long tradition of physical planning. During the Middle Ages experts (primarily hydraulic technicians, surveyors and master builders) were already grappling with various problems dealing with the layout and management of areas in the Netherlands - which was relatively densely populated even then. In the first half of this century civil engineers were still very involved in drawing up plans for urban expansion. Furthermore, in this period the influence of architects began to increase. The architects rejected the primacy of functionality in city architecture and demanded that more attention be paid to beauty, monumentality and the picturesque in the townscape (Van der Valk, 1983). This period also saw recreation gradually becoming an issue for government planning. One of the ways this was manifested was the CIAM (Congrès Internationaux d'Architecture Moderne) notion, formulated in the Charter of Athens (1933), in which the four main functions of urban development plans (housing, work, traffic and leisure) were segregated geographically (Beckers, 1983).

The 1920s and 1930s were characterized by growing government interference with many areas of social life. The legislation on urban development (the Housing Acts of 1901 and 1921) formed the starting point for preparing plans for urban extension. Regional plans (streekplannen) were introduced in the early 1930s. By the end of that decade there was a clear need for a National Plan to solve the most pressing problems in urban and rural development, and so in 1941 a national agency (Rijksdienst voor het Nationale Plan) was set up to produce a national landuse plan. Beckers (1983, 188 ff.) notes that from the outset this national agency requested that due attention be paid to outdoor recreation. It is noteworthy that even then recreation was seen as a phenomenon requiring spatial solutions, because of its massive character.

During the 1930s a new group appeared on the physical planning scene: social geographers from the so-called Amsterdam and Utrecht school. As a result, the role of the civil engineers gradually faded into the background: "So the engineers abdicated their research role to the scientists as they had previously lost their design role to the architects. They remained on the scene, however, to design (and later plan) infrastructural works, not the least important task in the Netherlands! Meanwhile, social researchers and architectural designers entered into a new struggle for leadership" (Faludi and De Ruijter, 1985, 44).

The social geographers of the 1930s laid the foundations for the strong social science accents in post-war physical planning in the Netherlands. Although this influence has by no means disappeared, since the 1980s there has gradually been more of an interdisciplinary approach in physical planning. Not only spatial, but also procedural and organizational matters have become of interest to planners.

5.2   The period since World War II

5.2.1   General

After the war the great appreciation of nature and the pursuit of sobriety, simplicity and independence that characterized Dutch recreation planning in the 1920s and 1930s gave way to the pursuit of luxury and comfort combined with a desire for freedom in the sense of informality (Hessels, 1973). Initially, private enterprise led the way in the further development of recreation, but in the 1960s and 1970s government influence became increasingly significant. The mid-1970s saw the demise of the 'redistribution mentality' and the government also began to focus more on development and growth. To foster the latter, the government began to work to create optimal conditions for market forces (Volkers, 1989). The government and non-profit organizations thus began to become more aware of how to apply principles of marketing. In the 1980s there was a switch from thinking in terms of supplying demand to market-focused thinking.

5.2.2   The institutional framework

In the Netherlands physical planning* is considered to be a matter of national policy. "Physical planning is not considered to be a policy in its own right but rather the center of one special sector of government concern. This sector deals with all the spatial aspects and all the spatial consequences of government management as a whole" (Held and Visser, 1984). "In the Netherlands physical planning has developed into a policy, which by proper coordination of the different policy

---

* For an overview see Held et al., 1984

sectors aims to ensure that from a spatial point of view these sectors are coherent and that they jointly contribute to the desired physical development" (Witsen, 1972, 102). Dutch planning consists of sector planning and aspect planning. A sector is a specific branch of government activity (housing, agriculture, tourism, for example). Sector planning refers to the concrete programming of various dimensions of a government activity. An aspect refers to a single aspect (physical planning, for instance) of different phenomena. Therefore, physical planning is strongly directed towards integrating the spatial aspects of various economic, social, recreational and other activities.

The Dutch physical planning system exists at three levels. At the national level there are overall plans, with a broad perspective. At the provincial level the regional plan is the most important instrument of physical planning. It roughly indicates the future development of a province. In a certain sense the regional plan is the starting point for physical planning at the lowest level of local government. At this local level the allocation plan has far-reaching consequences for all kinds of local activities. "The municipalities have the power (and for the area outside the built-up area even the obligation) to establish allocation plans. These plans indicate the allocation of the land and contain regulations as to the purpose for which the land and the buildings on it may be used. This plan is the only physical plan which is binding upon the citizen: a building permit is only granted if the project concerned is in agreement with the allocation plan" (Witsen, 1977, 101). Other plans are important too, particularly the structure plan. The latter gives the broad outlines of the future scenario envisaged for all or part of a local government area. It is not binding. The land reallocation plans and the rural reconstruction programs are two special forms of physical planning in rural areas.

The field of leisure can be divided into specific 'sectors' such as outdoor recreation, tourism, sports, culture, and media. As in most of the countries these different sectors are not integrated on the level of national policy. Different ministries are involved in outlining national strategies for sport development, planning of outdoor recreation facilities or stimulating economic development in tourist regions. Therefore a short overview of this complicated intermeshed whole seems to be appropriate[*].

Outdoor recreation

The Ministry of Agriculture, Nature Management and Fisheries is responsible for the outdoor recreation policy in the Netherlands. Within the Directorate Rural Areas

---

[*]   Mainly based upon Koot, 1990

72

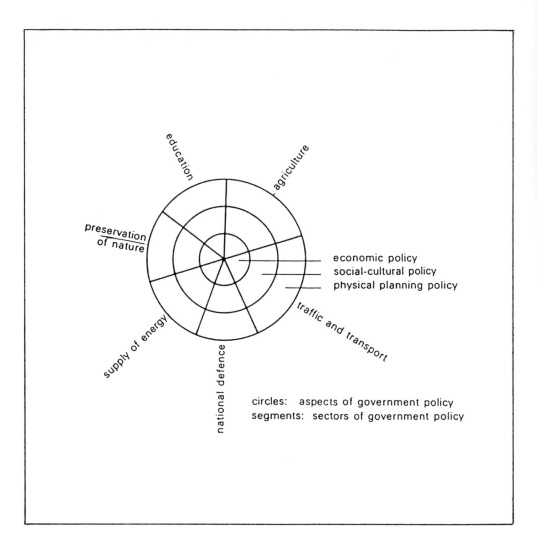

Figure 5.1   SCHEMATIC REPRESENTATION OF THE RELATIONSHIP
BETWEEN ASPECTS AND SECTORS OF GOVERNMENTAL
POLICY

and Quality Care the Directorate Outdoor Recreation has to care for the execution
of the national policy in this respect. The objective of this policy can be described
as 'the improvement of lasting availability of outdoor recreation possibilities for the
society'. Koot (1990, 12) specifies this objective as follows:

- Improvement and strengthening of the outdoor recreation sector in stimulating 'professionalism' by the improvement of knowledge and expertise of recreationists and suppliers of facilities
- the diminishing of shortcomings in outdoor recreation facilities
- the improvement of the quality of the recreation environment and the promotion of attainability and diversity.

Based on this specified objective the Directorate Outdoor Recreation has four main tasks (Koot, 1990, 13):

- A subsidizing task to stimulate investments for the development and maintenance of recreation facilities.
- A coordination task. The efforts of the different ministries with aspects of outdoor recreation are coordinated in the Coordination Commission Outdoor Recreation
- To maintain relation with organizations and associations in field policy advising and consumer interests
- Stimulating research on problems related to outdoor recreation. This research is carried out in specialized research institutions financed by the Directorate Science and Technology.

Tourism

The Ministry of Economic Affairs has the primary responsibility for tourism. Within the Ministry the department of Tourism is concerned with the policy performance, that is to stimulate tourism development in general and to support the private sector. Special attention is paid to marketing and promotion activities for which the National Agency for Tourism is responsible. The role of the Ministry can be described as creating conditions in favor of the private sector (Koot, 1990, 18).

By means of the Governmental Five Year Program on Outdoor Recreation and Tourism the financial instruments of the Ministry of Agriculture, Nature Management and Fisheries and the Ministry of Economic Affairs are combined.

Sports, culture and general leisure policy

Also the Ministry of Welfare, Public Health and Culture is represented on the national leisure scene. Especially sports and culture are cared for by this Ministry, although there are quite a lot of sport branches which are fitted in another Ministry (such as for instance water sports, equestrian sports and sport fishery).

74

Besides the Ministries just mentioned there are others involved in specific aspects of leisure activities. Important in this respect is the Ministry of Housing, Physical Planning and Environmental Control, because this Ministry is particular responsible for the spatial order or spatial configuration of the Netherlands, and its physical planning. As mentioned, the Dutch have a very tradition in physical planning and the influence of the various governmental Notes and Reports on spatial policy of the last three decades is remarkably strong.

On the local level municipalities play an important role in implementing the objectives of national recreation and tourism policy. Municipalities are responsible for the management of public green facilities, forests and landscapes. Indirectly, they have a great influence on the implementation of physical planning objectives through the allocation plans. In order to coordinate outdoor recreation policy on a sub-provincial level, different municipalities work together. A great number of so-called Recreatieschappen (Designated Outdoor Recreation Areas or DOR Areas) have been established based upon the Communal Regulations Act. The responsibilities of these DOR Areas differ according to size and complexity of the territorium involved. A DOR Area may be just one facility or a lot of facilities within a large area. A main objective of a DOR Area is the development of a regional outdoor recreation plan, the executing of this plan and the managing of the existing facilities for walking, cycling, swimming, windsurfing and fishing. At this moment 70 DOR Areas exist in the Netherlands.

5.2.3   Developments in policy and planning after World War II

In the years immediately after World War II Dutch policy on recreation retained a clear educational dimension. Emphasis was laid on the character-forming benefits of youth work, on improving the masses, and on camping. Recreation was recognized as a societal issue worthy of serious consideration. The significance of the spatial implications of recreational behavior increased, too. Outdoor recreation became a mass movement focused on relaxation. In the 1960s the educational aspect faded into the background and outdoor recreation came to be seen primarily as a problem of physical planning.

The development of physical planning in the Netherlands underwent an important change with the appearance of the First Report on Physical Planning in the Netherlands, which included an announcement of impending legislation on physical planning. The Physical Planning Bill was enacted in 1965. To be sure, this Act invested local government with far-reaching powers via the allocation (or landuse) plan, but it also gave the national government a task in the formulating of a coherent and consistent national planning policy. This spatial policy was prepared by the National Physical Planning Agency, the successor to the government agency for the national plan.

The First Structural Outline for the Spatial Development of Outdoor Recreation appeared in 1964. It was the result of implementing the policy lines laid down in the Report on Physical Planning. In this Outline the Netherlands was divided into areas to be developed primarily for day recreation and areas designated primarily for overnight recreation. A striking feature was the proposal for eleven 'large-scale elements for day recreation' to be constructed in the vicinity of the big cities. The 'work with work philosophy' plays an important role in the construction of large-scale amenities for recreation in the Netherlands (Spee and Van der Voet, 1989). Examples include the planting of woods in the Wieringer lake area and the Northeast polder, the way the Rand lakes have been laid out as part of the reclamation of the former Zuiderzee, and many of the hydraulic engineering works carried out in the Delta area of southwest Netherlands.

The Second Report on Physical Planning, published in 1966, was far more important than its predecessor. It was shaped by the assumption that the population in the Netherlands would grow rapidly, reaching 20 million in 2000. A central theme in this report was to achieve a more balanced population distribution, to relieve the high population pressure in the Randstad area. A hierarchically arranged system of parks for day recreation was designed: city parks, parks in the metropolitan district, and parks and watersport areas of regional and national importance. It is striking that overnight recreation and, in a certain sense, water sport, disappeared from the field of view of national and provincial government, whereas the recreational provisions of the cities were seen as a primary responsibility of local government.

5.2.4 Broadening the scope of recreation policy

The appearance of the Second Report on Physical Planning marked the beginning of the second phase. This period lasted until the 1970s and was characterized by an expansion of the policy area of outdoor recreation. The spatial aspect became more important, as did the socio-cultural aspect and, even more so, the natural environment (Spee and Van der Voet, 1989). With the creation of a new Ministry of Culture, Recreation and Social Work in 1965 the institutionalization of recreation at the national level became a fact. This new ministry developed its own tools to steer developments in recreation. Among the most important were the Procedure for Project Development in Outdoor Recreation (which appeared in 1969) and, at the local level, the agencies for cooperation between local government areas (the outdoor recreation authorities). Recreation policy had become a policy for providing amenities.

In the early 1970s it became clear that the recreational amenities that had been provided were not without problems. They were usually monofunctional and were also far from population centers. Furthermore, in the rural areas developments in agriculture (i.e. rationalization) and in the transport infrastructure, reduced the possibilities for day recreation. It became clear that spatially accommodating the developments in recreation was meeting more opposition from people concerned with conserving nature and landscapes, and also that in many places people felt that the local way of life was being threatened.

The change in the attitude to environment, nature and landscape was clearly visible in the Orientation Report (1973), which was the first part of the Third Report on Physical Planning in which a set of national goals and policy choices was proposed. This report contained the important decision to expose policy goals to public debate. The so-called 'crucial physical planning decisions' were to be reached after a long process of participation involving the various pressure groups.

The second part of the Third Report on Physical Planning (the Urbanization Report) did not pay excessive attention to the importance of recreation and tourism to the urban environment. The policy for urban development was only discussed in relation to the immediate environment. The section on the layout of the inner city did not discuss the tourism or recreation significance of the public areas in the inner city. In the policy recreation was still seen as an aspect of the rural area.

Recreation did have a clear place in the third part of the Third Report on Physical Planning (the Report on Rural Areas), which appeared in 1977. The national policy for rural areas had two thrusts:

* the policy in relation to the construction of large recreation areas ('green stars') was continued, but in a much diluted form;
* multiple use of land was stressed.

Within the framework of the Third Report on Physical Planning two Reports have to be mentioned as important for the development of outdoor recreation: The Structure Vision Outdoor Recreation and the Structure Scheme Outdoor Recreation.

The Structure Vision is aimed at formulating the sector policy on a national level. A structure scheme deals with the physically relevant aspects of a sector policy. "It is a report with maps on the long term policy to be conducted in regard to certain--mostly infrastructural--provisions, which are relevant for the physical planning policy and for which the government bears a large measure of responsibility" (Meijer, 1977, 152). The main objective of the Structure Vision is: "Promotion of a variety of outdoor recreation forms to enable free participation in principle by all categories of society. Forms of outdoor recreation that:

- contribute to self expression of the human being and the experience of nature and culture;
- correspond with the countries' physical characteristics in a quantitative as well as qualitative way;
- are managed such that they contribute to the quality of society." (Koot, 1990, 27).

Not until the 1970s did the awareness that rural areas were much more than areas of agricultural production emerge in the Netherlands. This was much later than in Britain, where in 1968 the Countryside Act pioneered the idea of multiple land use as a planning feature (Van Oort, 1989, 134). In the Netherlands the Land Consolidation Act of 1954 had as its primary objective the improvement of agricultural production. Planning had to precede the incorporation of natural and recreation elements in the countryside. The Land Development Act of 1985 explicitly sanctioned the multiple use of rural areas.

The spatial implications of the Structure Vision Outdoor Recreation and of the Report on Rural Areas were elaborated in the Structural Scheme on Outdoor Recreation, published in 1981, which put forward the policy proposals to be submitted to public debate. Based on the public debate and the advice of several organizations and interest groups the central government takes a decision to be submitted to the parliament. In the definitive version, published in 1985 and covering the period 1985-2010, the main goal is to offer all categories of the population the possibility of participating in various types of outdoor recreation. Concrete landuse planning measures for outdoor recreation are specified (Department of Outdoor Recreation, 1989). Concrete objectives of the Structure Scheme are:

- the performance of outdoor recreation facilities within the urban areas;
- the improvement of the inner city recreation facilities;
- the optimization of recreational use of rural and/or nature areas (multifunctional land use).

The emphasis in this policy outline is on eliminating the deprivation situation in urban areas. Furthermore in some areas the outdoor recreation development is restricted to specified zones.

Important also is the proposal of putting forward a specific green belt policy. This was implemented in the Report Physical Framework Randstad Green Structure planning of 1985. This plan can be described as performing a entity of green areas within the urban influence sphere of the Randstad to safeguard the rural area (the Green Heart of the Randstad) and to structure the development of urban functions in this region. It is obvious that the Randstad Green Structure planning is of great importance for the implementation of outdoor recreation facilities of different kind. In this planning framework the central government, the provinces and the municipalities coordinate their efforts.

More sectoral in character than the Reports and Notes just presented is the so-called Note Recreationists Policy published in 1986. The emphasis lies upon themes such as recreationists guidance, education and research, or otherwise formulated on the interests of the recreationist him or herself. The central government distinguishes a series of bottlenecks for each recreation activity. Special attention has been paid to deprived groups in the society (for instance ethnic minorities, unemployed people and disabled persons). The policy is put forward into four main lines:

- Strengthening the policy focussed on diversification, accessibility and attainability of the recreation possibilities and improving the quality and affordability of the facilities offered;
- Strengthening the policy referring to research, promotion, information and education;
- More attention to be paid to the role of advising organizations; and
- Strengthening the policy in management-legal sense.

In the Note Recreationists Policy concrete measures are proposed for specific leisure activities (including less known activities as terrain driving, air sport, keeping small animals and gardening). In a certain sense this Note marks the transition to another phase in recreation policy and planning in the Netherlands, reflecting the spirit of that time. It is more client-oriented (or market-oriented) but at the same time it has the characteristics of the welfare policy period of the late seventies in defining on the central governmental level the kind of deprivation and the deprivated groups.

5.2.5   New Realism: emphasizing the role of the market

In the early 1980s the conviction grew that physical planning was no longer the concern of one independently operating government. The pursuit of decentral-ization of political decision-making and the desire to allow all pressure groups to have a say in the drawing up of landuse plans led to the emergence of 'negotiative planning'. "In any case the Dutch are motivated by the best of democratic intentions, which form a solid base for this method of planning sometimes labeled 'communicative planning' in the vein of Habermas's critical theory. 'Negotiative planning' is another term of great currency in the Netherlands, having fewer overtones and encompassing the many dealings between various departments of government, rather than focusing solely on the relations between government and citizens" (Faludi and De Ruyter, 1985, 46).

Van der Cammen (1982) pointed out that planning was increasingly resembling a shared production process in which, at certain moments, various bodies meet round a table in order to negotiate agreements. A give-and-take mechanism arose within a relatively closed official circuit. In the mid-1980s this was how the Structural Outlines, the regional plans and the basic plans for outdoor recreation were prepared.

Yet this method of planning has not remained free of problems. Objectives formulated for the long term are still present--relics of the 'blueprint model' of planning. Recreation planning is still strongly oriented towards providing amenities and the number of amenities to be provided is calculated on the basis of supposed shortcomings. The consensus about the extent of the shortages and the decisions about how many amenities should be provided, and where, are reached within the official circuits, in the context of professionalized planning. The social consensus about the formulated aims is also gradually diminishing (Van Straten, 1984). The point at issue is not so much to attain an optimal landuse plan, but to attain a compromise.

Van de Cammen (1982) added another element to this criticism. He observed that in the early 1980s, in spite of the efforts made at the negotiating table, the landuse plan hardly held its own in the midst of sectoral planning. Physical planning has come to mean the antithesis to the interests of sectors, and the result is no more than a derivative of the highest common factor of the sector interests (Hidding and Kleefmann, 1989).

National policy for tourism and recreation in this period is strongly inspired by social and cultural developments, and consequently takes account of the effects of social processes in changing recreationers' preferred leisure behavior and wishes. A major objective of this policy is the development of a high quality recreation environment accessible to all. Another characteristic of this period is the predominance of the economic perspective and the strong market-orientation. Furthermore, the reappraisal of urban areas as a leisure environment has brought about a discussion on the leisure resources of inner cities, or urban elements, such as historical urban morphology, urban architecture, museums and shopping areas (Ashworth and De Haan, 1985; Jansen-Verbeke, 1988). The evaluation of these urban characteristics as resources for urban tourism and the recognition of their recreational function is one example of how the policies for tourism and for recreation have converged towards a common interest (Jansen-Verbeke and Dietvorst, 1987). The privatization of the management and exploitation of recreational facilities is another example of this trend. This makes it all the more surprising that a distinction is made between tourism and outdoor recreation at the level of national policy.

At regional and local level, however, outdoor recreation and tourism are intermeshed, partly because of the introduction of TROPs (Tourism Recreation Overall Plans). The creation of these plans has been stimulated by the Reports on Tourism Policy issued by the Ministry of Economic Affairs. The first such report was published in 1979 and the second in 1984.

In these Reports much attention has been given to the potential economic benefits of tourist development. Ashworth and Bergsma (1987) notify the large adverse balance of international tourist payments as a priority problem. This is to be mitigated by the stimulation of four aspects of the tourist industry (Ashworth and Bergsma, 1987, 154):

- The capturing of a larger share of high-spending intercontinental tourist trade to Western Europe;
- The diversion of part of the Dutch holiday market from foreign to domestic; destinations;
- The encouragement of 'near-neighbor holiday-making' especially from the West German market; and,
- A more profitable exploitation of European transit tourists.

Ashworth and Bergsma emphasize the limited possibilities for the implementation of the just described policy. Some measures in the sphere of infrastructure and the promotion of "The Netherlands- A country of water" can be mentioned, but they point out the fragmentation of control within the tourist industry, the divided responsibility for different aspects of tourism between different government ministries and the autonomous character of the marketing strategies of large hotel and leisure companies as interfering to a successful tourism policy.

For a long time tourism has been regarded as well suited to bringing about economic prosperity in peripheral regions. An important instrument for this is the TROP. So far, twelve regional TROPs, based on a synthesis of the results of investigations into regional tourism resources and recreational amenities, have been proposed. Some criticism has been leveled at these plans, particularly at the way priorities between recreation and tourism have been handled, how these plans relate to other political instruments of planning, and how they are intended to create a link between the authorities and private enterprise (Bouman and Lengkeek, 1986). Jansen-Verbeke and Dietvorst (1987) criticized this new trend in physical planning as being the result of an 'ad hoc' reaction and not of a well-balanced rethink--a trend that is even more obvious in the present development of many local plans. Commercial motives are currently tending to override many other concerns in physical planning.

TABLE 5.1   DUTCH MARKET SHARE IN INTERNATIONAL TOURIST
EXPENDITURES 1980-1988

Dutch share	1980	1981	1982	1983	1984	1985	1986	1987	1988
Europe	3.3	3.4	4.0	3.8	4.2	4.1	4.3	4.3	4.3
N-W Europe	6.3	6.4	7.6	6.9	7.4	7.0	7.9	8.0	7.8

Source: Nota Ondernemen in Toerisme, 1990, 13

TABLE 5.2   TOURIST RECREATION EXPENDITURES AND EMPLOYMENT
IN 1989

	Expenditures (in billions)	Direct employment (in men year)
National day recreation and holidays (including short breaks	12.2	72,000
Foreign expenditures	6.4	37,000
Mediation (tourist agencies)	6.7	14,000
Mediation (business travel)	3.7	12,000
Tourist related goods	2.0	4,000
Total	31.0	139,000
Indirect employment		38,500
Total employment direct and indirect		177,500

Source: Nota Ondernemen in Toerisme, 1990, 17

In 1990 the Ministry of Economic Affairs published a third Report on Tourism called 'Enterprising in Tourism'. It emphasizes the growing importance of tourism for the Dutch economy. Expenditures in tourism and recreation rose by 33% between 1982 and 1989. Now circa 10% of the Dutch economy is entirely or at least to a large degree dependent on tourism and recreation. The central government objective for tourism policy is to stimulate a sustainable growth. Not only within the framework of restricting environmental and nature conditions but also in making natural and cultural resources more profitable.

It is evident from table 5.2 that the share of Dutch tourist expenditures is dominant. The tourist and recreation related expenditures amount to 9% of the Dutch consumption expenditures. Day trips and holidays account for 20.9 billion guilders, an amazing amount compared with the total expenditures for clothing and shoes, for instance.

The national policy on tourism for the coming period put forward in the Report 'Enterprising in Tourism' is based upon a SWOT-analysis of the Dutch tourism sector. The following table summarizes the main points of this SWOT-analysis.

After years of preparation and discussion a new Governmental Policy Report on Outdoor Recreation was published at the end of 1991. The national government itself regards this Report as a real trend rupture in outdoor recreation policy. The report put forward a selective attitude. The national government retreats from direct intervention and sees the provision of favourable conditions and opportunities as a primary goal. The new policy puts emphasis in four issues:

- Recreation in a natural environment
- The Netherlands, A country of water
- Recreation and urbanization
- Strengthening of the recreation sector

For the implementation of this policy a selection of recreation tourist regions is made, which has however no relation with the spatial differentiation laid down in the Fourth Report on Physical Planning. But this is not the only critical remark. The new Report was received with severe criticism from almost every group within the "world" of recreation. Actually the responsible department in the Ministry of Agriculture, Nature Conservation and Fisheries was struck by severe budget cuts. Inside the Ministry the traditional agricultural interests dominated those of outdoor recreation. The Advisory Council for Physical Planning, amongst others, reacted very critically on the new Report: 'A poverish political content, the marginalisation of this policy area and insufficient financial means', was the conclusion.

The positive impact of the bad receipt of the Report was the starting of a debate on the position of outdoor recreation in the National Policy. In this debate a plea is made for long term investment in recreation and tourism, and public authorities are challenged to take their responsibilities in assuming the quality of the tourist recreation space.

TABLE 5.3   A SWOT-ANALYSIS OF THE NETHERLANDS AS A TOURIST RECREATION PRODUCT

	STRONG/TREATMENT	WEAK/THREATS
GENERAL	The relation between the Netherlands and water	Significance not enough recognized in society
	Rich natural-historic heritage	Insufficient quality
	The coast	Young tourist sector not yet professionalized
	Spatial concentration of the tourist recreation product	Great dependence on automobility
	Congress tourism	Long procedures in planning location or extension
	Huge potential market caused by growing holiday participation in Southern Europe	Procedural problem in product adjustment Diminishing knowledge of foreign languages
	Potential destination for soft tourism	Problems of pollution and social security in big cities
	Interesting destination for short-break holidays	Growing international competitiveness
		Not enough attention paid to market segmentation
		Insufficient public transport facilities for TR-concentration areas
INDUSTRY	Promising growth opportunities	Not enough professionality
	Elaborated classification systems	Low education level of labor force
	Strong congress product	Fragmented structure and at random growth of education institutions
	High quality in hotel management education	Low degree of automatization
	Growing cooperation between line organizations	Insufficient developed (marketing) information systems
CONSUMERS	Asking for higher quality	Not enough attention for distribution of information and client complaints
	High Dutch participation rates for holidays and low participation in foreign countries	Unbalanced delivery conditions in travel arrangements Insufficient international coordination with respect to third-party risks
	Growing market for seniors	Absence of international agreements on classification systems and guiding designs
	Growing cooperation between industry and consumer interest groups/ organizations	
INFRA-STRUCTURE	Existing infrastructure of high quality	Insufficient use of European funds
	Pursuing the correction to private investments	Inability of local authorities to choose strong points for tourist development
	Growing cooperation between the different levels of public policy	Insufficient opportunities for multifunctional use of rural and urban regions
	Still growing interest for tourism development by local authorities	Lack of market orientation in product development

Based upon a survey of strong and weak points presented in the Report 'Ondernemen in toerisme', 1990

5.2.6 Stimulative planning: challenge for the nineties

The impact of strong market-orientation and of the importance of economic goals in physical planning was already apparent in the Fourth Report on Physical Planning published in 1988. That report reaffirmed the 1960s concept of physical planning at the national level - i.e. global planning or 'stimulative planning': "A major shift of emphasis is that the Fourth Report anticipates change more by adjustments to quality than by increases in quantity. In addition, the emphasis lies much more than in the past on reinforcing strong points and on exploiting opportunities which offer potential. Spatial policy must help bring about favorable conditions and help avert threats" (Ministry of Housing, Physical Planning and Environment, 1988). The Fourth Report sees the basic characteristics of spatial development in the 1990s as being

1.   the urban intersections at which a regional clustering of amenities can be created;
2.   economically strong regions benefiting from the idea of regional autonomy;
3.   an increased interest in the development of the economic core area, the so-called Central Netherlands Urban Ring; and,
4.   the launching of the concept 'The Netherlands--A country of water' to reinforce the quality of the wet environment and to improve facilities for water recreation.

5.3   Conclusion

At the end of the 1980s a call was made to strengthen the integration function of the landuse plan. In physical planning (and hence also in planning for tourism and recreation, because the spatial element is so important) the opportunity for outlining possible and desired spatial elements must be seized. Plans for tourism and recreation land use should attract public attention once more; they should stimulate the imagination and provoke public debate. In short, planning for tourism and recreation should be stimulative planning.

There is another important argument for stimulative planning. In the 1970s and 1980s recreation planning was governed by the supply mentality. Attempts were made to outline the quantitative need for recreational amenities. These quantitative prognoses were not very reliable; furthermore, the supply mentality led to a static 'tourism and recreation product'. During the 1980s it was realized that public tastes had changed and that the amenities for outdoor recreation in many ways no longer satisfied the demand. It was realized that the image of strong normative planning which focused on achieving an 'ideal' situation was no longer adequate (Van Straten, 1984). The need for more flexible and more market-oriented forms of physical planning was becoming increasingly apparent. The recent debate on the latest Report on Outdoor Recreation gives hope for the implementation of such forms of planning.

LITERATURE

Ashworth, G. and Th. de Haan, 1985. The touristic historic city: a model and application in Norwich, Field Studies Series, no. 10 Geographical Institute State University Groningen, Groningen.

Asworth, G.J. and J.R. Bergsma, 1987. New policies for tourism: opportunities or problems, Tijdschr. v. Econ. en Soc. Geografie, 78(2): 151-155.

Beckers, Th. Planning voor vrijheid, 1983. Een historisch-sociologische studie van de overheidsinterventie in rekreatie en vrije tijd, Dissertatie Wageningen.

Bouman, A. and J. Lengkeek, 1986. Een vergelijking van toeristisch-recreatieve ontwikkelings-plannen, Vrije Tijd en Samenleving, Vol. 4(3): 10-15.

Cammen, H. van der, 1988. Environmental planning in the Netherlands in the 21st century. In: L.H. van Wijngaarden-Bakker and J.J.M. van der Meer, Spatial sciences, research in progress, Nederlandse Geografische Studies 80, Amsterdam, 84-90.

Cammen, H. van der, 1982. Methodisch geleide planvorming, Stedebouw en Volkshuisvesting, 377 385 and 449-459.

Department of Outdoor Recreation of the Ministry of Agriculture and Fisheries, 1989. The Dutch Government policy on outdoor recreation, The Hague.

Dietvorst, A. and M.C. Jansen-Verbeke, 1988. De binnenstad: kader van een sociaal perpetuum mobile. Nederlandse Geografische Studies 61, Amsterdam/Nijmegen.

Faludi, A. and P. de Ruijter, 1985. No match for the present crisis? The theoretical and institutional framework for Dutch planning. In: Ashok K. Dutt and Frank J. Costa (eds.), Public planning in the Netherlands, Oxford University Press, Oxford, 35-49.

Held, R. Burnell and D.W. Visser, 1984. Rural land uses and planning. A comparative study of the Netherlands and the United States, Elsevier, Amsterdam.

Hessels, A., 1973. Vakantie en vakantiebesteding sinds de eeuwwisseling, Assen.

Hidding, M.C. and F. Kleefmann, 1989. Het facetbegrip in de ruimtelijke planning, Stedebouw en Volkshuisvesting (4): 38-41.

Koot, H.E., 1990. Outdoor recreation policy in the Netherlands and Great Britain. Analysis and comparison of the key players and the main aspects that determine outdoor recreation policy in the Netherlands and Great Britain, Leiden.

Jansen-Verbeke, M.C., 1988. Leisure, recreation and tourism in inner cities: explorative case-studies, Nederlandse Geografische Studies, 58, Amsterdam/Nijmegen.

Jansen-Verbeke, M.C. and A. Dietvorst, 1987. Leisure, Recreation, Tourism: A geographic view on integration, Annals of Tourism Research, Vol. 14(3): 361-375.

Meijer, P.G. Structure schemes for infrastructure. In: Planning and development in the Netherlands, Vol. IX(2), Van Gorcum, Assen, 149-168.

Ministry of Housing, Physical Planning and Environment, 1988. On the road to 2015.Comprehensive summary of the Fourth Report on Physical Planning in the Netherlands, SDU publishers, The Hague.

Ministerie van Economische Zaken, 1990. Nota Ondernemen in Toerisme, Den Haag.

86

Nieuwenkamp, J., 1985. National physical planning: origins, evolution and current objectives. In: Ashok K. Dutt and Frank J. Costa (eds.), Public planning in the Netherlands, Oxford University Press, Oxford, 72-86.

Oort, G. van. The application of the concept of multiple use of land in the Netherlands: a comparative case study, 1989. In: G. Clark, P. Huigen and F. Thissen, Planning and the future of the countryside: Great Britain and the Netherlands, Nederlandse Geografische Studies, 92, Amsterdam, 134-144.

Spee, R.J.A.P. and J.L.M. der Voet. Inleiding Recreatiekunde, Werkgroep recreatie.

Straten, A. van, 1984. Recreatieplanning in onzekerheid? Een probleemverkenning. In: A. van Straten (ed.), Recreatieplanning in onzekerheid, verslag van de studiedag op 15 juni 1984. Werkgroep Recreatie, Landbouwuniversiteit Wageningen.

Valk, A. van der, 1983. Opleiding in opbouw. Geschiedenis van het Planologisch en Demografisch Instituut van de Universiteit van Amsterdam 1962-1982, Planologisch en Demografisch Instituut van de Universiteit van Amsterdam, Amsterdam.

Vierde nota over de ruimtelijke ordening, deel a: beleidsvoornemen, 1988, Den Haag.

Volkers, C.R., 1989. Perspektieven voor landelijke gebieden. In: P.P.P. Huigen and M.C.H.M. van der Velden (eds.), De achterkant van verstedelijkt Nederland. De positie en funktie van landelijke gebieden in de Nederlandse samenleving, Nederlandse Geografische Studies 89, Amsterdam/Utrecht, 131-146.

Witsen, J., 1977. Crucial physical planning decisions In: Planning and development in the Netherlands, Vol. IX(2), Van Gorcum, Assen, 99-114.

CHAPTER 6

Planning for tourism and recreation: a market-oriented approach

by
Adri Dietvorst
Professor and Chairman Center for Recreation Studies
Dept. of Physical Planning and Rural Development, Agricultural University;
Head, Dept. of Outdoor Recreation and Tourism
The Winand Staring Centre for Integrated Land, Soil and Water Research;
Wageningen, The Netherlands

6.1   Introduction

Characteristic for the development of tourism and outdoor recreation in the eighties is the shift from 'supply-thinking' to 'demand-thinking'. Suppliers of all kind of tourist and recreation goods and facilities focus upon the needs and preferences of potential users and/or visitors. Consumers are more critical than before and look for quality, variety and challenge. The market for leisure goods and services in general has become fragmented and therefore a new approach in tourist recreation planning and product development is necessary.

The ongoing differentiation in the demand for leisure goods and services is a manifestation of a typical characteristic of (post)modern society: VARIETY. At least three aspects of variety are significant for the planning of tourism and recreation:

- The variety in the types of households. The composition of households in Western countries is more differentiated than ever before;
- The variety in leisure behavior within the households as a consequence of emancipation and individualization; and,
- The variety in leisure needs and preferences of each individual. The need for constant alternating of stimuli and activities (hopping) is becoming increasingly apparent.

These aspects of variety influence the leisure-related use of space and time. Let us therefore first examine the major changes in the demand from tourists and recreationers in the 1980s and then elaborate the consequences for planning and product development.

6.2   The process of ongoing differentiation

For a long time it was expected that technological innovation and growing prosperity would lead to a general increase in leisure time. But developments in the Netherlands have been otherwise. The working week has been shortened somewhat (from 40 hours to 38 hours), but the goal of 32 hours, which was to be reached by 1992, is no longer realistic. However, there has been an important increase in part-time employment.

The reduction of working time implies an increase in free time (i.e. time free from job-bound obligations) but not an increase in leisure time, as studies done by the Social and Cultural Planning Office (SCP) have made clear.   Extensive time-budget analyses have revealed that leisure time has decreased slightly, because there has been an increase in travel time, more time being spent on training and because falling incomes have encouraged people to spend more time doing their own home improvements or repairs instead of employing tradesmen. Only   economically inactive persons (especially the elderly) appear to have had an increase in their leisure time in the period 1975-1985 (SCP, 1988). "The idea that free time is continuously increasing is very popular. If the Dutch population could actually avail themselves of the time that they have variously been credited with, then they would have little time over for work, household chores, study or sleep. Furthermore, it is unquestioningly assumed that those who display an active free-time behavior or who are always open to new challenges are also those who receive more time for such things" (SCP, 1988, 219).

The graph shows that free time is not increasing at the same rate for everybody. Persons who use free time actively have not experienced an increase in net free time. (Those who make use of a relatively large variety of free-time facilities are, on average, younger, better educated and have a relatively high consuming ability.) The Socio-cultural Report notes that: "largely because in the 1980s the groups that are generally receptive to new activities were busier with their daily tasks it seems that they have little flexibility in their pattern of free time" (SCP, 1988, 225). That means that if a new activity is taken up it almost certainly replaces another. "It is probable that the competition for the free hour will therefore heat up".

6.3   Demographic processes

Demographic trends suggest that the demand for all types of accommodation and for leisure facilities will generally increase in the coming years. Although population growth fell from 1.27% in 1970 to 0.38% in 1983, there has been an

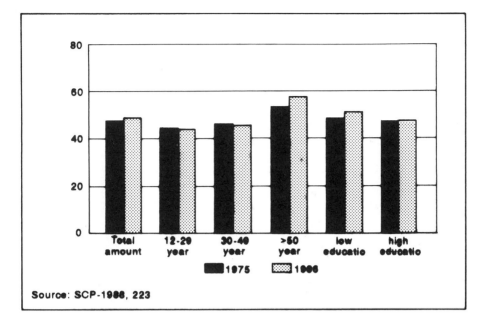

Figure 6.1   INCREASE IN NET FREE TIME BETWEEN 1975 AND 1985
(HOURS P.W.)

upturn since then. In 1990 the growth percentage was 0,79%. There is a remarkable spatial differentiation in the average population in the last 15 years.

The <u>absolute increase</u> in population is leading to a growing demand for various amenities for recreation and tourism, although it can be assumed that participation in vacations (currently 75%) is unlikely to grow much more in the coming period. Hardly any further growth is expected in summer vacations in the Netherlands, but additional vacations and short trips are expected to increase (Dijkhuis-Potgieser, 1988).

6.3.1 Maturing

The maturing of the population, that is, the relative decrease in the number of young people, will continue. It is associated with the enormous change in the number of births, a process that has already lasted 25 years. In 1970 in the Netherlands there were still 18 births per 1000 inhabitants; this figure has fallen to around 12. Various reasons have been put forward for this fall: secularization and the decline in church-going, the emancipation of women and their resulting willingness to participate in work, and a lifestyle centered on the prosperity of the individual. Nevertheless it cannot be concluded that young people no longer have an

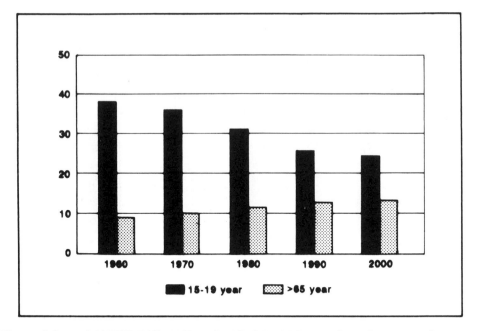

Figure 6.2   MATURING AND AGING OF THE DUTCH POPULATION

important role in how free time is spent. Although their numbers are falling they do influence the picture in certain areas--for example, in cities. The quantitative decline in young people is therefore also spatially differentiated. Moreover, young people cannot be considered to be a homogeneous group. Five types can be distinguished, on the basis of lifestyle (an amalgam of values, attitudes and activities). Approximately 39% of young people are bourgeois, 15% are critical of society, 10% are self-centered, 30% are autonomous and 15% are career- and consumption-oriented (the total percentage exceeds 100 because there is overlap in this typology). From this typology, which is considered to be representative of young people in the Netherlands between 12 and 24 years of age (in total, 3, 159, 377), it can be inferred that an appreciable number (40%) are unlikely to change their behavior drastically. After all, the bourgeois types follow more or less conventional patterns of behavior. For the other groups it is probable that the more general social tendency towards individualization and commercialization will become very evident. Moreover, it is safe to assume that young people living in the inner city are more likely to belong to the 'autonomous' or 'critical of society' group than young people living in the countryside. And these are the very groups that can be expected to stimulate change that can also be important for the inner city area (Dietvorst and Jansen-Verbeke, 1988).

6.3.2   Aging

In a certain sense, aging is the counterpart of maturing. The aging process is possibly one of the most radical social processes that will confront our society in the coming decades. This demographic development will lead to relatively more older people and fewer younger people to look after them. Two sorts of aging can be distinguished: the absolute and the relative, and the two reinforce each other.

Absolute aging is to do with the falling death rate, which is associated with a change in life expectancy. In 1845 the average man in the Netherlands lived for 36 years and the average woman for 38 years. These figures are now 74 and 80 years, respectively. The relative number of old people also has increased, because of the falling birth rate (the maturing of the population). This process of relative aging has accelerated considerably since the 1960s. The factors that have contributed to this are largely social and cultural, such as better level of education, the emancipation of women (and the use of contraceptives) and increasing affluence. The traditional age of retirement, 65, is often taken as the lower limit for the third phase of life. According to the medium prognosis of the Central Bureau of Statistics the number of people aged 65 and above in the Netherlands will increase from 1.8 million to reach 2.2 million by the year 2000, at which stage 14.5% of the population will belong to this group. The process of change will be accelerated in the first few decades of the 21st century by the post-war baby boom generation, so that by 2030 approximately one-third of the population in the Netherlands will belong to the 'third age group' (Van der Wijst and Van Poppel, 1985).

One result of the demographic aging is that society will age, with all that this implies for the layout and management of town and country (De Gans, 1986, 249). Social aging denotes a society in which the behavior patterns, social interests and priorities will largely be determined by and for the third age group. What makes the aging process an extremely interesting social process for planners is that it occurs gradually. Its effects can be determined beforehand, because the senior citizens of the year 2000 are already among us. The behavior pattern of the middle-aged group is probably a strong indication of the behavior of senior citizens in the future.

Proceeding from the assumption that a future generation of senior citizens will behave significantly differently from the present generation of senior citizens, I believe that more attention should be paid to the implications of this for leisure-time activities in general and for the tourism and recreation sector in particular.

Senior citizens by no means form a homogeneous group either. The group is strongly differentiated according to education, income status, age and ability. It can be expected that senior citizens will show more diversity in their requirements for leisure time than they do at the moment. Even though social attitudes tend to be constant there are some signs of change: there are new developments especially in sport, fitness training and health therapies of all sorts. The increased interest in health therapy possibly will be stronger in the coming generation of senior citizens. The trend is now mainly visible in better educated and higher income groups, but it can be expected to spread across other groups in future.

Some changes might also occur in the patterns of away-from-home activities. The present middle-aged groups have a greater 'awareness space' and, moreover, have a relatively better level of education and greater mobility (they own automobiles and hold driving licenses), and this can lead to a more varied use of time (including leisure time) which is continued into a later stage of life too.

6.3.3   The new households

The changes in the composition of the Dutch household are just as important for the development of recreation and tourism as the maturing and aging of the population.

Since the war the composition of households in the Netherlands has become much more differentiated. Although the traditional family (working husband, housewife with two or more children at home) is by no means in the minority, deviating types of households (single persons, double-income couples, single-parent families) are growing in importance. This group is generally referred to as 'new households'. The rise of new households is the result of a complex of demographic, socio-economic and socio-cultural factors which include changes in the birth rate, the differences in life expectancy between men and women, a decrease in the number of marriages, a relative increase in the number of divorces, and the erosion of the traditional role-bound distribution of tasks between men and women (RPD, 1986).

Van Engelsdorp Gastelaars (1989) observed that the Netherlands is currently experiencing developments that go hand in hand with an increase in social differentiation. Households in the Netherlands are becoming increasingly diverse in terms of their daily allocation of tasks, their incomes and the leisure time available to them. This pattern of differentiation is expected to continue in the coming years. This means that a continued growth in the diversity of lifestyle and the recreation pattern of the various categories of inhabitants can be expected. Van Engelsdorp Gastelaars distinguishes four categories of household:

1.   The traditional autochtonous family
2.   The starters
3.   Single workers
4.   Older couples with income from sources other than work.

1. The traditional autochthonous family

This is characterized by the asymmetric division of tasks between husband and wife. In general, the man is the breadwinner and the wife looks after the house and the children. Leisure time is mainly spent together: "This is largely spent around the home (watching TV, drinking tea, gardening) and much less elsewhere (visiting relatives and friends, societies, sport, cycling, short trips). Hence the accent is largely on pursuits with a social character, that is, with a strong emphasis on being with family, friends, team-mates and others.

Because a large proportion of the household budget is spent on living expenses and other fixed costs, the amount of income available per family member for additional activities is not very large. This means that leisure time activities must not (indeed, cannot) cost much. Regarding living locations, this type of household is primarily found in the suburbs (in single-family dwellings with a garden, in neighborhoods with a relatively low density of houses, plenty of grass and room for children to play, scope for walks and cycle trips.

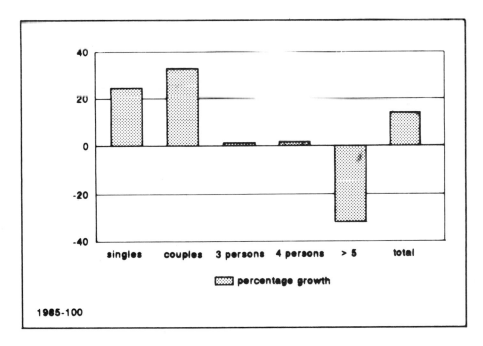

Figure 6.3   CHANGES IN THE COMPOSITION OF HOUSEHOLDS 1985-1995

2. The starters

This group is characterized by being relatively well educated but not having a very high income. The fixed income (from study grants or social welfare benefits) is supplemented by occasional temporary employment. Van Engelsdorp Gastelaars finds that this group typically spends little time in formal administered work. There is plenty of time left for other activities: "Young starters work as decorators and mechanics, or selling second-hand clothing and books, or producing printed matter or furniture, or as waiters or cleaners. Many of these activities take place in the 'informal' economy."

These strategies, and the proportionally lower living expenses, mean that persons in this group have a fairly high proportion of their income available for leisure pursuits. This group can therefore spend relatively large amounts on all sorts of free time activities: "More than other types of households, starters spend large amounts of their free time away from home, partly in cafés, theaters, discotheques, sports halls, music shops, and partly on the street, in open-air concerts, cycling or jogging in parks, demonstrating or in pressure groups". They do not have enough money to travel very far, hence their preference for living in inner cities.

3. Single workers

This type of household developed in the 1970s. In general, these are households with no children. The members of the household are well educated, participate actively in the work process, have a relatively high income and are aged between 30-50. They are typified by the strategies they adopt to monetarize and automate the household chores so that they maximize their leisure time. This time-saving strategy yields them almost as much leisure time as the adults in a traditional family. But they spend their free time very differently: Their recreation style is appreciably more oriented outside the home. They spend much time in public places, such as restaurants, theaters, museums, or outdoors (jogging, cycling, playing tennis). This type of household is likely to be found living in neighborhoods in or near city centers.

4. Older couples with income from sources other than work

These are people in the age group 55-65. According to Van Engelsdorp Gastelaars they are typically households in a reduction phase: Any children have grown up and left home. Usually, the couple are retired and therefore their income is not very high but they have plenty of free time. Sometimes this free time is spent on money-saving pursuits (gardening, repairs, mending their clothes).

Their leisure pursuits take time and are cheap, like those of unemployed persons. It has been found that rather than adopting new activities, these people spend more time on their old hobbies or pursuits. From research done in Amsterdam, Van Engelsdorp Gastelaars found that many of the couples in this group had a second 'home': "an allotment garden with a summerhouse, a caravan or a canvas home in a camping site, a boat with bunks, or, if more cash is available, a real second house. These people, especially those who live in towns, spend large parts of the year in and around these types of 'dwelling'.

6.4 Recreation styles

This sketch of the demographic and socio-cultural variety in the Netherlands will suffice for the moment, although it is possible to elaborate it further. For example, to point out the variant ways of spending leisure time current among the ethnic minorities (a quantitatively not unimportant group). Given the constraints of this essay, I will close this overview of the growing differentiation in lifestyles in the Netherlands with a short description of the variability in recreation styles. A typology of tourists will be given in section 6.8.2.

In his research on the effects of privatization in the Meuse lakes area of the Dutch province of Limburg, Philipsen (1989) distinguished four types of water sport enthusiasts. People who camp on farms can also be classified into four groups, according to the degree to which they choose this form of camping on principle, and the degree to which they appreciate comfort. These groups are: the principled type, the comfort-oriented type, the complex type and the opportunist (Zonneveld, 1988).

The consequence of the differentiation in household types can be perceived in the 'recreation styles' put forward in a study done by Andersson and De Jong (1987) in two adjoining neighborhoods in Rotterdam. Lengkeek (1989) comments that some influences of the size of the household budget and free time could be traced, but that there were also indications that the differences and changes in recreation behavior could better be explained by lifestyle routines and patterns, as well as by the dynamics in lifestyle characteristics.

Lengkeek summarizes the distinction Andersson and De Jong (1987) made in recreation styles as follows:

1. 'In search of myself'; these are young adults, living on their own, trying to find a suitable lifestyle and a professional basis; networks of friends are important for leisure activities. Leisure activities must not be very expensive

2.   'Life with no prospects'; young unemployed adults with few skills; time is spent passively; outdoor recreation is of minor importance

3.   'How to find time for myself'; young couples, both working; considerable household income; time is scarce; high consumption level; outdoor recreation means a break with daily routine and being together

4.   'As long as the kids don't go short of anything'; working-class families with one wage-earner; life is centered around home and family; outdoor recreation is spent within a small radius (cycling, walking)

5.   'We manage to get along'; the older working-class couple, now retired with low income; leisure is concentrated in and around the house; allotment gardens, fishing and a stroll in the park are activities that structure daily life

6.   'Enjoying retirement', well-to-do retired couples; outdoor recreation is important to keep fit and to structure daily life.

6.5   Spatial consequences of the differentiation processes

The diversity in the tourism and recreation sector is made visible by a continuing differentiation in demand. For the suppliers in this sector this means an increase in the number of market segments and the need to have to take account of the very divergent wishes and expectations of the different users' groups. The consequence is a more client-focused approach and a great diversification of the tourism and recreation product. In fact, in the 1980s the tourism and recreation market changed from a supply market to a demand market.

If we restrict ourselves here to the visible geographical effects of the strongly differentiated spatial behaviour of tourists and recreationers, we see that this leads to previously separated phenomena and/or areas becoming more involved with each other. In other words, the process of social differentiation is leading to spatial integration and thus also to an intermeshing of functions. However, in the rural areas the pressure to separate functions has by no means been removed. There, a combination of processes has led to a separation of functions and to the creation of isolated recreation amenities, which Van der Voet (1989) called "recreational flowerpots". The future countryside of Europe is in danger of becoming characterized by "extensive industrial agro-landscapes and displaced and derelict residual landscapes" (Regt, 1989), as a result of processes such as the scaling up of agriculture. In this context Blaas (1989) sketched a somber picture for the rural areas in the Netherlands if their development were to be left to what he called 'exogenous forces': They would degenerate into areas for holding special events, accommodating transient visitors and for retreating.

The mono-functionality in parts of the urban residential area and the loss of diversity in the rural area therefore go against the grain of social processes such as individualization, differentiation in lifestyles and changing behavior in time and space. These processes manifest themselves in the tourism and recreation sector by the presence of a motley variety of recreational activities and recreational styles. This variety is one of the reasons for the specialization among the suppliers of tourism and recreation goods, services and amenities. But specialization cannot continue indefinitely, because limits are put on the base on which tourism and recreation amenities rest. A dilemma looms, because a desired specialization or differentiation cannot be realized, for lack of interest. Small, specialized but nevertheless essential amenities can disappear in this way. In the market sector this happens simply because companies fold up; in the public sector this happens by amenities being neglected, being shut down, or being disposed of by privatization.

6.6   The need to create tourism and recreation complexes

A solution could be to counterbalance the tendency for specialization among the suppliers of tourism and recreation facilities by creating <u>tourism and recreation complexes</u> (Dietvorst, 1989). These complexes are characterized by not having their elements spread randomly over an area. To the visitor the amenities appear to be related to each other and therefore need to be near to each other. Thus, the whole is seen as more attractive than each separate amenity. This is a spatially differentiated whole. Hence, tourism and recreation phenomena need to be regulated locally, regionally, nationally, and sometimes even internationally. 'Regulating' means ensuring that the elements of a tourism and recreation complex assume certain positions relative to each other. This regulation means not only considering what the tourism and recreation product has to offer, but also how the visitors use various aspects offered by the tourism and recreation complex. These positions of the product elements distinguished can be described as subordinate, coordinate, attracting, reinforcing, and complementary. These designations indicate that there is some sort of association--spatial as well as functional.

This implies that the development of our tourism and recreation product needs to be tackled in a different way. There is a need for well structured thought about the product in the market sector and, even more so, in the sector of subsidized outdoor recreation. It goes without saying that government and industry must in most cases develop an integral vision of the total product. However, we have not yet gotten that far. In many local TROPs tourism and recreation areas or towns are seen as a simple addition of attractions, facilities and supporting infrastructure. This is aggregation rather than integration, let alone the creation of complexes.

When creating tourism and recreation complexes a consensus must be reached about a large number of conflicting interests. We must therefore take account of a large diversity in interest groups which, in our democratic society, have previously undreamt of access to large amounts of information. This brings us up against the not inconsiderable problems ensuing from the information explosion of recent decades. Huppes (1987) has indicated the consequences of over-information in our society. The vast amount of information available means a greater chance that the information one person selects will differ greatly from that selected by another, and that the proposals shared by all parties concerned will have less influence on decisions.

6.7   Stimulative planning

One possible solution for the development of a tourism and recreation complex which does justice to the diversity in interests and in insights, is to use the opportunities offered by modern insights in the possible applications of physical planning. Kleefmann (1984) indicated the possibility of using physical planning as an instrument for discussing as yet unrealized directions of development. Physical planning then becomes an instrument for orienting the well-considered presentation of target scenarios. This is called stimulative planning (Boelens, 1988; Ashworth and Voogd, 1988).

At a national scale this turnabout in planning research has recently been demonstrated in the presentation of the New Netherlands project (a private initiative to develop four scenarios for 2050) and also to some extent in the Fourth Report on Physical Planning. Van der Cammen (1988) comments on these as follows: "What the two plans, which have just been mentioned, have in common is that they are not explicitly and primarily meant to be carried out in the future, at least not in the usual meaning of that word. First of all, the function of the designs for 'Nieuw Nederland' is to draw the public's attention to the importance of improving our country's environment as a long-term affair. The Fourth Report, if one reads it well, chiefly has a stimulating and initiating purpose also".

The concepts of the significance of stimulative planning have also penetrated the local level. Some of the new-style structure plans (Comijs, 1989) were also intended to promote discussion about the ideas presented. This is not just the regulation of the spatial effects of social processes: it also triggers new ideas and activities. Various examples of this can be seen in the Fourth Report on Physical Planning. Raggers and Voogd (1989) follow Wissink (1987) in pointing out that in this sense planning is more than determining a course: "it also serves to give the relationship to the action to be taken". They propose three conditions for this form of physical planning:

1.   When drawing up the plan the market processes must be made visible. Therefore there must be conditions that enable demand and supply to be dynamically attuned. That also means that all parties must be involved in the content of the plan. Raggers and Voogd also note that making the market processes visible does not necessarily mean following market developments blindly, and in this context they make a useful distinction between market-conforming planning (where this occurs) and market-oriented planning.

2.   Spatial priorities must be 'clearly specified'. This setting of priorities sends a signal to users and potential users.

3.   It must be assumed that this signal works as an incentive. It must lead to action.

Raggers and Voogd also point out that there must be an open planning process. This would enable a brainstorming session to be organized for the many interested parties at an early stage. They work this out further by using a real example: the open planning process for the Verbindingskanaal zone in the city of Groningen. The difficulty of providing sufficient counterweight to the strong emphasis that is laid on the economic function in many urban renewal processes strengthens the argument for identifying the networks of interests and for finding solutions to guarantee the greatest possible internal variety. The priority is not to exclude certain groups, or to reduce the complex structure to a few dominant variables (e.g. economic), but to maintain the greatest possible diversity and hence the maximum dynamics.

A central dilemma here is the problem of maintaining (in this case, improving) landscape or urban qualities in the long term and, in the short term, the desired flexibility for the market-oriented action within the tourism and recreation sector. The business community must be flexible, in order to be able to keep its position in the market; local inhabitants sometimes have short-term interests. When giving more form to stimulative planning a way must also be found to reconcile demands for quality that have their origin in processes with totally different time courses. Failure to do this results in developments like those currently visible in the European Alps or, to take an urban example, in Venice. Framework planning (cascoplanning) might solve this dilemma. In this approach "a distinction is made between functions that require stability and continuity and functions that benefit from flexibility and adaptability. This could lead to a spatial framework (casco) which accommodates the functions with low dynamics, and to 'built-in assembly kits' for functions with high process dynamics. The casco requires long-term planning with a delineated end goal; the functions with high dynamics can be regulated via term planning" (Bakker et al., 1989).

Having outlined the significance of complexes and networks for the development of the tourism and recreation sector I now propose to consider the consequences for research and planning.

6.8   Complexes and planning perspectives: ideas for implementation

In an extremely instructive paper presented at the ELRA congress--'Cities for the future'--Ashworth and Jansen-Verbeke give a conceptual framework for the identification and analysis of urban tourism and recreation complexes. In it the emphasis is on the spatial element within the functional associations that are present. Their argument stems from the need to pay more attention to how the various elements of a tourism and recreation complex are integrated mutually and with other urban functions.

Their ideas form the starting point for the development of a planning strategy that does justice to the principles of stimulative planning as well as to those of strategic planning. I shall deal with these planning strategies one by one.

6.8.1   Mapping the product elements

The first step in the planning process is mapping the various product elements at a specification appropriate to the level of analysis. Our hypothetical planning region measures 15 km x 15 km. For the inventory an appropriate typology of tourist or visitor attractions needs to be available. Such typologies have been made from various viewpoints. Lew (1987) reviewed the research methods used in the study of tourist attractions and distinguished three broad perspectives: the ideographic definition and description of types of attraction, the organization and development of attractions, and the cognitive perception and experience of tourist attractions by different groups. A modified ideographic typology will serve our purposes; the organizational and the cognitive perspectives can be saved for the next steps in the planning process.

Lew based his 'Composite Ideograph Tourist Attraction Typology' on the distinction between attractions that are nature-oriented and those with a human orientation. However, he used the criterion of generality to some extent in distinguishing 'general environment', 'specific features' and 'inclusive environments'. Despite being aware of the problem of classifying infrastructure and service facilities, Lew did not adequately solve this problem. The fact is that visitors use these facilities, but they are not necessarily attracted to a specific location to see them. Therefore, I propose to follow the solution proposed by Jansen-Verbeke (1988). In her study of the inner city as a tourism and recreation product she defines this area as a set of primary, secondary and additional elements. This basic

TABLE 6.1   THE CLASSIFICATION OF THE ELEMENTS OF A TOURISM
            AND RECREATION COMPLEX

NATURE	NATURE-HUMAN INTERFACE	HUMAN
PRIMARY ELEMENTS		
1. General environment Panoramas Climate Landscape Nature reserves National Parks	Observational Rural Botanic Gardens Archeology	- Settlement morphology - Monuments - Canals/harbors
2. Specific features Landmarks	Leisure Nature Trails Parks Beaches Resorts Hunting grounds	Socio-cultural factors - Liveliness of a place or region - Regional culture - Folklore
3. Inclusive Environments	Participatory All kinds of outdoor activities	Cultural and/or amusement facilities - Museums - Theaters/cinemas - Festivals - Fun fairs - Cuisine
SECONDARY ELEMENTS		
Hotels Restaurants Bungalow parks	Camping sites Shopping facilities Rental facilities	Markets Cottages Picnic sites
ADDITIONAL ELEMENTS		
Accessibility and parking facilities Tourist facilities, Information offices, signposts Guidebooks, Maps		

classification criterion could also be relevant for rural areas of tourist regions that are at a larger scale than the inner city. As a consequence one has to identify the primary elements, that is, those characteristic elements and/or amenities displaying a 'magnetic' function for tourists and visitors. These primary elements refer to places where people can be active (theater, swimming pool, beach, boulevard, organized events), to the physical leisure setting (for instance, the historic landmarks of a town, the parks and green areas, the forest area) or to the socio-cultural sphere (folklore, local customs and the like). Jansen-Verbeke defines the secondary elements as those not exerting an autonomous attraction on visitors but nevertheless representing an important aspect of the attractiveness of a place. These secondary elements include hotels, restaurants and cafés, the markets, the camping sites, and the rental agencies (for boats, bicycles or automobiles). She considers these secondary elements as very important in terms of the amounts of time and money spent.

Finally, there are additional elements. They do not attract visitors to a specific location (a national park or a museum) but they are more or less necessary conditions for a pleasant visit and thus for the functioning of a tourism and recreation complex. This category includes parking facilities, information offices, signposts and the availability of adequate maps and guidebooks.

The combination of the ideographic typology constructed by Lew and the distinction Jansen-Verbeke made referring to the authenticity of the attraction of a specific element is presented in table 6.1.

6.8.2 Creating complexes out of spatial clusters

After the inventory work the product elements can be amalgamated in clusters, following a specific criterion of spatial association. To a certain extent this clustering is intuitive, demanding much local or regional knowledge. A detailed analysis of the tourist recreation product in a specific region generally reveals spatial concentrations of various product elements. This is due to the presence of one or more dominant production factors. The dominating factor 'beach' brings about a spatial concentration of various product elements such as road infra-structure, restaurants, hotels, campsites, mobile catering (ice, hot dogs, french fries, beverages) and facilities for sports and games. In the historic city of Enkhuizen, to mention another example, a concentration of elements as restaurants, historic monuments, yacht harbours, museums, green structures, shopping center and tourist information service occur in a relatively small area. If nothing else is done than pure description of the spatial concentration of existing product elements, and much tourist recreation planning does not reach further, we prefer to use the concept 'cluster' in determining the concentration of elements. The concept 'complex' has to be reserved for sets of tourist recreation product elements with a specific coherence. In determining the presence of this coherence, this functional association, two different procedures can be followed.

First the complexes can be determined according to their organizational, legal and financial aspects. Referring to this, the following questions are important:

- To what extent the relationship between the product elements can be described in terms of competition, complementarity or dependency?
- What is the relation between product elements of the public sector and those belonging to the private sector?
- Is there any cooperation regarding marketing research and promotion?
- To what extent 'foreign' enterprises (hotel chains for instance or multinational leisure companies) exercise control in the local tourist economy?
- Does the regional or local tourist economy have at its disposal networks as driving forces for economic growth?

The just mentioned questions can be extended by others, but in a period in which so much attention is asked for the capriciousness of the market, I prefer to start a client-oriented procedure.

One of the most crucial phases in the planning process proposed here is the identification of tourism and recreation complexes as experienced by potential and actual visitors. It is essential to determine the relationship between the complexity of the spatial context and the more individual leisure experiences. So tourist recreation complexes have to be created or identified in analyzing the coherence within and between clusters of product elements from the viewpoint of the visitor. It is the visitor, the tourist and/or recreationist who make the tourist recreation complex through a specific time-space behavior. Different target groups (based upon different recreation or leisure styles) form different tourist recreation complexes (TR-complexes). It is obvious that spatial scale plays an important role in determining the different complexes. A particular product element can belong to a variety of tourist recreation complexes. The famous Kröller Muller Museum in the National Park 'Hoge Veluwe' belongs to the TR-complex created by Japanese tourists visiting the cultural capitals of Western Europe (Paris, Amsterdam, Kröller Muller Museum, Frankfurt). But, it forms also an important element for some tourists staying at a campsite in the surroundings of Otterlo, a village 6 km from the Kröller Muller Museum. In the latter case the complex could be composed of: the campsite, the cycle roads on the Veluwe, the shopping facilities in the nearby city of Arnhem, the restaurants in Otterlo and so on.

It is also evident for this client-oriented procedure to make a classification or typology of target groups. Again two different strategies can be put forward each resulting in a specific research methodology:

- By means of desk research in setting up a typology of target groups. This can be done by matching the product elements supposed to be used by a specific target group with the product elements present in a region. The wishes of day-trippers will differ from those of campers, and walking tourists will want different facilities than water recreationers (see table 6.2 for one of the possible tourist typologies). This procedure will give a first global impression of the tourism and recreation complexes present in a city or a region. The case study Ede to be presented in section 6.8.3 is a demonstration of this procedure.

- Through empirical research in exploring the time-space behavior of tourist and/or recreationists. Empirical research based upon the time-space behavior of recreationers and tourists is effective, though difficult to execute. The theoretical concepts developed by Hägerstrand and other researchers from the so-called Lund school might be helpful. For the moment, however, we have to cope with a considerable lack of empirical data on the time-space behavior of visitors. Some experience has been gained in measuring the multiplier effects of visiting the local museums in the Dutch cities of Arnhem and Nijmegen (Dietvorst et al., 1987; Ten Tuynte and Dietvorst, 1988). Visitors were asked to send back a questionnaire on which they had recorded their 'time-path' in the city before and after the visit to one of the museums involved. Shopping and using outdoor cafés and restaurants ranked high among the preferred activities. The appreciation of open spaces and waterfronts just for walking and sightseeing was also noteworthy. In section 6.8.4 we discuss the first results of empirical research carried out in Enkhuizen. Interesting insights into the actual patterns of use and users of inner urban areas are also reported by Jansen-Verbeke (1988). She did a survey in the inner areas of three medium-sized towns in the east of the Netherlands to analyze the urban functions from the users' point of view. The results were systematically summarized in a series of visitor profiles, pointing out how a specific group differs from the average population (Jansen-Verbeke, 1988, 190).

6.8.3   Case study Ede

In 1989 local authorities of Ede, a municipality in the province of Gelderland, asked for the identification of tourist recreation complexes to be used as a starting point for a Structure Vision on the development of tourism and recreation. Because lack of financial means necessary to carry out an empirical survey, the research group decided to use the desk procedure (SBW, 1990).

After the inventory of tourist recreation facilities in the municipal area, ten clusters of product elements could be distinguished. Each cluster has a hierarchical structure, conforming with the classification in primary, secondary and additional elements proposed by Jansen-Verbeke (1988). To give an impression of the resulting clusters, three of them will be described shortly:

TABLE 6.2   TYPOLOGY OF TOURISTS

---

1. The adventurous and active vacationer
   - Always on the lookout for new experiences, seeks adventure and the unknown. Always making new plans. Prefers to go his own way.

2. The sightseeing and museum freak
   - Is very interested in historical places. Likes to visit museums and old churches. Interested in folklore and local culture.

3. The sociable vacationer
   - Wants to relax, prefers to go off with a group. Sets high store on personal contact during the vacation.

4. The lazy laid-back vacationer
   - Wants to be left alone to laze about. Occasionally indulges in sport.

5. The home-from-home vacationer
   - Prefers a familiar environment, not too far from home. Goes with the family. Walks and cycles a lot. Looks for a certain type of accommodation (bungalow).

6. The weekend and short break vacationer
   - Likes to get away for a few days regularly, but not too far away. Prefers a hotel or bungalow.

- 'Harskamp'. A small village with some campsites, some of them with facilities for long stay. Some of the surrounding forest areas are not accessible. The area west of the village has an agrarian character.

- 'Ginkel'. A heath area with facilities for daily outdoor recreation. Walking, cycling and horse-riding are possible. The cluster has parking facilities, picnic seats and some catering facilities. There are a few campsites.

- 'Bennekom'. A village with good shopping facilities and some restaurants in the village-center. East of the village is an extensive forest area with restaurants and campsites (chalets and immobile caravans for long stay). In the forest area, road bounded recreation is permitted. A very attractive small-scale rural area with riding schools and a swimming pool lies west of the village.

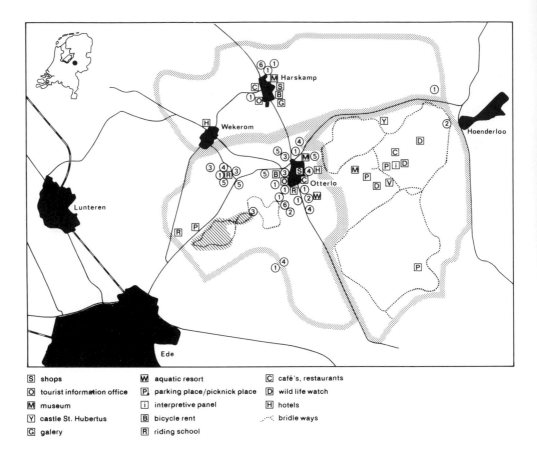

S	shops	W	aquatic resort	C	café's, restaurants
O	tourist information office	P	parking place/picknick place	D	wild life watch
M	museum	i	interpretive panel	H	hotels
Y	castle St. Hubertus	B	bicycle rent		bridle ways
G	galery	R	riding school		

Figure 6.4   SOME CLUSTERS IN THE MUNICIPALITY OF EDE

These facility clusters are used by a variety of target groups. Based on literature and short interviews with local experts ten types of recreationist/tourist are constructed including: 'the nature explorer', 'vacationers with a diverse activity pattern', 'vacationers looking for quietness', and 'cultural tourists'. For the first two mentioned target groups the description is given:

'Nature explorers'
Interests   Strong interest in nature and landscape qualities
Directed on other people   Low
Activity characteristic   Active
Type of recreationer   Day or staying visitors
Facility preferences   Low

The target group of nature explorers is attracted to the area by the ecological qualities. They cross the region by foot or bicycle. They are satisfied with simple accommodations and do not prefer much luxury. Small campsites with low facility level are preferred. Footpaths, cycle roads, shopping facilities in the small villages, some catering and/or restaurant facilities and the natural area are the elements in this tourist recreation complex.

'Vacationers with a diverse activity pattern'

Interests	Strong interest in nature, sports, culture, entertainment and shopping
Directed on other people	Low
Activity characteristic	Active and passive
Type of recreationer	Staying visitors
Facility preferences	Relative high

This target group consists of couples (eventually with children). Their activities are related to entertainment and therefore characterized by great variety.

In the 'Ede area' ten tourist recreation complexes have been constructed. Each complex has primary, secondary and additional elements. In figure 6.5 this can be seen for the cluster the 'Harskamp'.

After this desk research a SWOT-analysis (Strength-Weakness-Opportunities-Threats-Analysis) was started to bring about the essential factors for a product development strategy. This SWOT-analysis does not differ essentially from the SWOT-analysis connected with the empirical procedure and will therefore be discussed in section 6.8.5.

6.8.4 Case study Enkhuizen

Enkhuizen is one of the historic cities found around the former Zuiderzee, which is now Lake IJssel. It was granted city status in 1355 and enjoyed great prosperity during the 16th and 17th centuries, mainly due to herring fishery and to trading by the East and West India Companies. The city has many fine historic buildings, such as the 'Dromedaris', an old defensive tower at the entrance to the harbor, the ' Koepoort', entrance to the city from the land and several warehouses (some of them now used by the Zuiderzee museum).

The Zuiderzee museum (200,000 to 300,000 visitors annually) shows how an important part of the Dutch population lived and worked around 1900. This is done in the open-air part of the museum, where a complete village of 130 houses, workshops, shops and shipyards, complete with canals, roads and gardens gives an

CLUSTER: HARSKAMP	COMPLEXES									
	1	2	3	4	5	6	7	8	9	0
PRIMARY ELEMENTS										
Hotel/restaurant 'De Harskamp'	▓	▓	■		▓			▓	▓	
Campsites	░	■	■	■	■					
Chalet parks		■	■							
Camping on the farm	░			■						
Cycle roads in an attractive environment	▓	▓	▓	▓	▓			■		
Attractive roads for touring by car		▓	▓		▓			■		
SECONDARY ELEMENTS										
Camping on the farm	░	░	░							
Museum Infantery Schietkamp		▓	▓					▓	■	
Art galery "Zuid"									■	
Recreation-grounds near horeca-facilities		▓						▓		
Shops	░	▓	▓	▓	▓					
Long distance footpaths	■									
Trail cabin	░									
ADDITIONAL ELEMENTS										
Tourist Information Office		░	░	░						
Bicycle renting facilities	░	░	░	░						

Figure 6.5   THE TOURIST RECREATION COMPLEXES IN THE CLUSTER HARSKAMP

impression of the lifestyle of fisher folk around the old Zuiderzee.

The city has two camping areas: The 'Vest', and another, larger, area adjoining the Lake IJssel and near the Zuiderzee museum and the Fairytale Wonderland. The lake has good facilities for water sport, and in the summer Enkhuizen attracts many traditional wooden sailing ships. Not surprisingly, Enkhuizen is also a popular harbor for modern sailing vessels.

The following map gives a simplified impression of the tourism and recreation cluster in the Enkhuizen region. Not all the product elements are represented.

In the Enkhuizen case study the time-space behavior of visitors is of essential importance for the identification of tourist recreation complexes. The research project is not finished at this moment (1992), so the results given here are provisional.

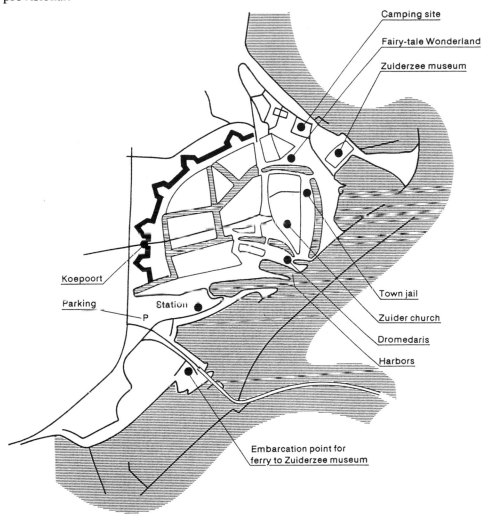

THE ENKHUIZEN TOURISM/RECREATION CLUSTER

Figure 6.6   THE ENKHUIZEN AREA

1. Time-space analysis of tourist behavior

When we consider man and his leisure behavior, we can no longer think about static structures and relationships. Rather we have to focus explicitly upon the dynamics of time-space relationships. Time-space relationships have a fairly long tradition in geographic research. Famous in this tradition and fundamental for the scientific development in this respect are the ideas put forward by Torsten Hägerstrand. Based upon spatial diffusion studies he tried to find concepts which might give integration to the basic dimensions of time and space. Since 1966 Hägerstrand and his research team at Lund (Sweden) have been working to elaborate these concepts.

The approach chosen by Hägerstrand and his followers can be considered as a physicalist approach to society (Thrift, 1977), an appreciation of the biophysical, ecological and locational realities which impose constraints on human activities. It is in fact a normative approach. Capability constraints (biologically based, for instance the need for sleep and food), coupling constraints (people have to do activities with others) and authority constraints (certain activities are controlled, not allowed or not possible at a given time) restrict daily activity patterns. Individuals are forced to pack their activities into specific time-space stocks. The emphasis is on constraints. On the other hand this 'constraint-oriented' approach is contrasted with the 'choice-oriented' approach (Floor, 1990) making use of time-budget analysis. Activities are considered as the results of choices. According to Chapin (1974) motives and preferences, time-space opportunities and time-space related context are influencing specific choices to be realized into concrete activities. In fact, the two mentioned approaches do not exclude each other and in practice a mixture would be most appropriate.

Basic assumptions in developing our time-space analysis of tourist behavior in Enkhuizen are (paraphrasing Thrift, 1977,6):

- The indivisibility of a human being. Time spent at a specific location can not be spent elsewhere at the same time;
- The limited availability of time to spend on a specific day;
- The fact that every activity has a duration and that movement between points in space consumes time; and,
- The limited packing capacity of space.

The way people make their choice within the constraints of time-space framework depends upon:

- Their motives, preferences and experience;
- Their images and estimations of opportunities; and,
- Their material resources.

To illustrate how different the time-space paths of individual tourists to Enkhuizen could be, given the basic assumptions just mentioned, we have constructed two hypothesized models.

2. Two hypothesized models of time-space behavior of tourists

Let us consider the 'water sport tourist' as our first potential customer for a tourism and recreation complex. For our first example we will consider a well-off well-educated German couple without children. Their hypothetical daily time-path starts in Enkhuizen harbor (in 1989 some 10% of the tourists came to Enkhuizen by boat). They take the small ferry from near the railway station, to visit the Zuiderzee museum. This museum is a primary element in the Enkhuizen tourism and recreation cluster. On the basis of a study done in 1989 it can be assumed that the visit will last about 3 hours. After visiting the museum the German couple decides to go for a walk in the town. They combine this sight-seeing walk with a meal in one of the restaurants. Back on their cruiser in the afternoon they decide to hire bikes and go off on a short cycle tour, returning in the early evening. The daily time-space path ends with a walk along the harbor front.

Our second example refers to a young couple with two children. Their time-space path on our hypothetical day starts at the camping site near Lake IJssel. They plan to visit the Fairytale Wonderland with their children. This visit takes about two hours. Before going to the town they return to the camping site for lunch. Then they walk to the town center to do some shopping. They do not go sightseeing in the historic parts of the city because the children do not show much interest in this. At four o'clock they return to the camping site and stay there for the rest of the day.

Of course, these examples could be extended by many others, to reflect the significant tourist types.

3. Methods for analyzing flow patterns and time-space behavior

Several methods can be used for analyzing the time-space behavior of the Enkhuizen visitors. We will briefly give an overview:
1. By using principal component analysis patterns, revealing 'visitor preference spaces' can be explored . The result of principal components analysis gives combinations of visited elements, and comparing these

112

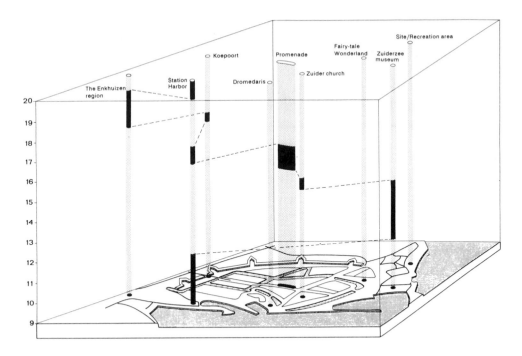

Figure 6.7   HYPOTHETICAL TIME-SPACE PATH OF A GERMAN COUPLE

patterns with visitor characteristics more insight is gained in existing tourist recreation complexes.

2.   By using network planning in combination with GIS (Geographic Information System). Real world conditions can be converted into models. GIS in land use planning gives the opportunity of adding possibilities for a translation of the real world into a model-world. Often operations research is used in modelling the maximization or minimization of the effects of certain human decisions. For the analysis of time-space behavior of visitors the routing solving capability of operations research might be very appropriate. A problem such as 'given a road infrastructure system, how to walk from point A to point B, if also points C,D and E have to be visited within a specific time-constraint', can be solved using operations research.

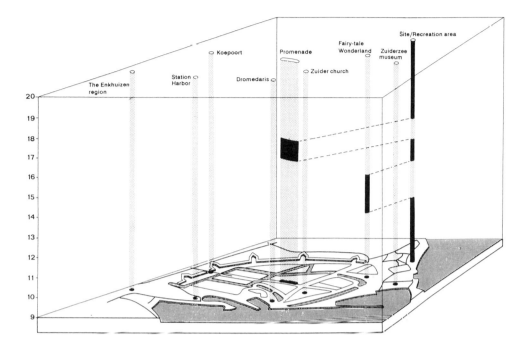

Figure 6.8   HYPOTHETICAL TIME SPACE PATH OF A YOUNG COUPLE
WITH CHILDREN

The computer program INTRANET (Jurgens, 1991) has been developed in solving planning questions related to land use planning and infrastructure. It is possible to grade the network in terms of accessibility, the type of land-use it has to serve, the economic qualities and even grading of a more psychological nature (for instance does the route offer enough variety) can be added. So INTRANET gives the opportunity to calculate several solutions in optimizing the coherence of a tourist recreation complex.

3.   Besides the application of operations research, networks can also be analyzed using mathematical graph theory. The classical example of this in geography is its application by Nystuen and Dacey in their analysis of telephone interaction to determine a regional hierarchy (Nystuen and Dacey,

114

1961). Their method is suitable to analyze flows of tourists in order to discover underlying hierarchical structures within tourist recreation complexes. The origin-destination matrix representing the network reflects the prevailing structure of linkages and dominance.

4.   If the researcher has the disposal of flow data for a sequence of periods, Markov chain analysis is very appropriate to describe the process of change within a tourist recreation complex (Dietvorst and Wever, 1977). The comparison of several calculated so-called transition matrices is extremely suitable in tracing the tendencies of change in the system observed. Unfortunately, however, one seldom has the opportunity to make use of these analyzing techniques because of lack of adequate data on a regional or a local level.

4. Some preliminary results of the Enkhuizen research project

The most striking result of the surveys held during the Easter period and the Summer period 1990 was the existence of at least three visitor groups according to their time-space behavior: The city-visitor, the Zuiderzee museum-visitor and what is called the combination-visitor (of both city and museum). Important are the duration differences of the core activities of the three groups just mentioned. The museum-visitor spends on the average 3-4 hours in the museum, enough reason to look away from a visit to the city. The poorly developed relation between the historic city and the Zuiderzee museum in the tourist-recreation complex 'Culture tourists' is strengthened by the results of the local access policy. Visitors to the 'open-air department' of the Zuiderzee museum have only access to the museum by using the ferry, and although they can leave the museum for a walk in the city they usually take the ferry back to the parking places located outside the city. 'Being pressed for time' and 'not planned' are often heard as motives for not bringing a visit to the city.

Only 44% of the museum visitors taking the ferry from the parking places outside the city, visit also the historic city, whereas 75% of the museum visitors taking the ferry from the departure near the railway station (thus more or less inside the city) also visit the historic city. Probably the visitor perceives the accessibility structure of the museum not in accordance with reality.

During the Summer survey tourists were asked which elements of the city were visited. The results have been analyzed by using principal component analysis revealing some 'visitor preference spaces'. We used principal components analysis because no particular assumption about the underlying structure of the variables is presumed. The PCA gives the best linear combination of variables (the best combination accounting for the variance in the data). Table 6.3 gives the results.

TABLE 6.3   PRINCIPAL   COMPONENTS   REVEALING   'VISITOR PREFERENCE SPACES'

	1	2	3	4	5	6
Townhall	.0256	.6765	.3473	-.0093	.0344	.0344
Jail	-.0759	.1036	.8005	-.0714	-.1444	.0287
Orphanage	.0531	.6257	.0083	-.1159	-.1777	-.019
Wester Church	.0920	.6290	-.0927	.0253	.4352	-.064
Zuider Church	.0542	.7095	-.0208	.0646	.3244	-.111
Fish-auction	.1836	.0142	.7201	-.0265	.2493	-.136
Koepoort	.1073	.1153	.5510	.1846	.5719	-.011
Drommedaris	.3761	.5115	.3612	.2081	.0333	.3794
Summer garden	.0492	.0545	.0627	.6762	.0385	.0527
Zuiderzee museum	-.7111	-.0185	.0833	-.1021	-.0698	.1815
Waag Museum	.0721	.2341	.0712	-.0495	.6367	.0935
Childrens pleasure	.0862	-.0639	-.0050	-.0817	.6683	.0535
Fairytale Wonderland	.0180	-.0853	-.0942	-.0495	.1197	.9099
Enkhuizen beach	.0583	-.0670	-.0330	.7365	-.0282	-.000
Shopping center	.7375	.0007	.0978	-.0269	.1160	.0492
Horeca	.6962	.2343	-.0828	.1172	.1388	.1914
Sailing Enkhuizen	.1761	-.0210	-.0612	.6949	-.090	-.095
Yacht harbors	.6919	.0473	.2377	.1965	-.033	.0333

The results presented in table 6.3 can be interpreted as follows:
- The first principal component representing 20% of explained variance, can be interpreted as 'City in general': shopping, visiting restaurants and sight-seeing in the harbors. These are high loading variables.
- The second component accounting for 11.3% of explained variance, can be pointed out as 'Historic scenery'. Historic monuments have the highest variable loadings.

- The third and fourth component can be called 'Local activities I' and 'Local activities II', because these components represent visits to events with an 'activity character' (for instance Fish-auction or Sailing Enkhuizen). Both components together account for 15.8% of explained variance.

The last two components 'Attractions for children I' and 'Attraction for children II' account for 12.3% of the embedded variance.

Unfortunately, the factor scores based upon the just mentioned PC-analysis revealed but a few interesting significant results, mainly due to the relative small number of respondents for each of the constructed target groups. The group with children could be related to the component 'Attraction for children II'. The sociable vacationer did not score on the component 'City in general', and the sightseeing and museum freak scored negative on component 'Local activities I'.

A visit to the 'open-air department' of the Zuiderzee museum is a time-consuming activity. Using network planning in combination with GIS, the INTRANET program was used to calculate alternative routes in the city of Enkhuizen for visitors to the Zuiderzee museum. The intention of presenting these models is to make clear that different time-space paths are possible and that the analysis of time-space behavior of tourists is extremely important in validating the strong and weak points or elements of a tourist recreation product in a given area. A detailed analysis of time-space behavior gives interesting insights for local politicians to strengthen the internal relation structure of the tourist recreation product.

Figures 6.9, 6.10 and 6.11 show the results. The first is based upon a presumed visit durance of three hours in the Zuiderzee museum, a visit to the inner city including a restaurant break accounting for two hours, and a rest time spent on a walk along the Drommedaris and the yacht harbor. In this model the tourist ends his visit to Enkhuizen at 16.00 at the railway station.

The alternative model is based upon a visit of two hours to the Zuiderzee museum, visits to some churches and monuments, a short visit to the Waag museum and a rest time to be spent for a walk along the Drommedaris and the yacht harbor.

Based upon the analysis of time-space behavior an 'Attraction/Duration port folio has been made (Sybrandy, 1990) (figure 6.12).

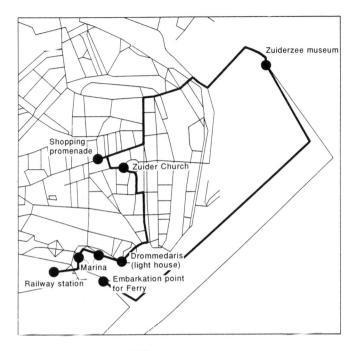

Figure 6.9   A PRESUMED VISIT TO FOUR SITES

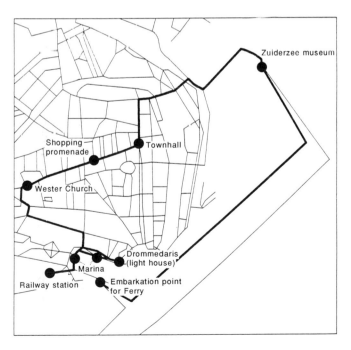

Figure 6.10   A PRESUMED VISIT TO SEVEN SITES

118

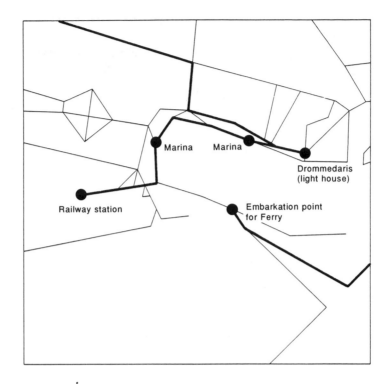

Figure 6.11   A PRESUMED ALTERNATIVE VISIT TO SEVERAL SITES

Core elements and symbolic elements are the primary attracting elements of the city. They differ in the possibilities for active use. A core element enables visitors to stay a fairly long time (the Zuiderzee museum or a restaurant), the symbolic element represents the image of the historic city (the Drommedaris tower). You must have seen these city marks but it does not take much time. Votary elements are visited by specific though not very large groups. They enjoy, for instance, the scenic beauty of a specific open space in the city, or in the case of Enkhuizen, prefer a long walk on the remnants of the city walls. Complementary elements are additional elements necessary for a visit (parking places, bank facilities, for example.)

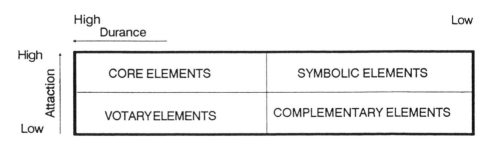

Figure 6.12   PORT FOLIO ATTRACTION/DURANCE

6.8.5 The SWOT analysis

Having ascertained the significant tourism and recreation complexes in a specific region, their qualitative strength (or weakness) needs to be known. One of the main objectives in managing tourism and recreation complexes at a regional level is to ensure the continuity of the tourist and/or recreation region as an attractive area for potential visitors. Ensuring continuity in terms of maintaining the market position or even improving the relative market share is an essential issue in the process of strategic planning. A well known procedure for this is the SWOT analysis (Strength-Weakness-Opportunities-Threats).

Many methods are available to study the SWOT elements of tourism and recreation complexes. Most of them originated in the context of marketing management at the levels of company, factory or corporation. Economic geographers have tried to transpose the typical management and/or marketing methods for corporate strategic planning (see for instance Kotler, 1988) to more spatially relevant methods (Ashworth and Voogd, 1987; De Smidt and Wever, 1984). The concepts of the 'product life cycle', of the 'positioning strategies', of Ansoff's 'product/market expansion grid' and finally of 'portfolio analysis' have proved to be useful in developing more spatially-oriented analytical tools.

For the analysis of the weak and strong points of entire tourism and recreation complexes in a specific region it would be appropriate to use a combination of product life cycle and portfolio analysis. Positioning strategies and/or the Ansoff model could be useful for determining the set of strategic choices for future development.

At a regional level the analysis starts with the application of the product life cycle concept. Cooper (1989) demonstrated the analytical strength of this method and Plog (1974) used the idea of the destination life cycle in combination with personality profiles. The results of Plog's research showed a product development in which each period of the life cycle has its own characteristic attraction for a specific market segment.

Obviously, a tourist region does not offer a single product but an abundance of product elements, each with a particular product life cycle. It would seem more appropriate to use portfolio analysis for a more detailed analysis of the qualitative status of the various product elements within a tourism recreation complex. Portfolio analysis can be applied to the region as a whole or to part of it; that is, it can be used for a comparative analysis of the competitive position of each of the tourism and recreation complexes in a region and also for an analysis of the weak and strong points within each of the complexes distinguished.

The original portfolio model as developed by the Boston Consulting Group shows a growth-share matrix, divided into four cells: question marks, stars, cash cows and dogs (Kotler, 1988). This model was criticized because the portfolio-matrix with four cells was considered to be too simple. More sophisticated portfolio-matrices consist of 9 or more cells, making a balanced judgement possible. The original factors such as market growth and market share are replaced by competitiveness and attractiveness. The ideal situation is one with a dynamic mix of activities or product-market combinations in a region or in a tourism and recreation complex. There has to be a balance between fast- and slow- growing 'product elements' (i.e. relatively few dogs and question marks and also a limited number of stars). The following figure is an illustration of a portfolio analysis of the product elements in the Enkhuizen area. The port folio is related to the primary product elements in Enkhuizen and based upon visitor figures and personal observations. Due to lack of detailed data (for instance on visitor expenditures at each product element) a more sophisticated port folio of Enkhuizen for a series of tourist recreation complexes could not be given at this moment.

6.8.6 Product development strategies

The final phase in the planning process for tourism and recreation complexes (in a specific region) is choosing certain product development strategies. Of course these strategies have to be created for different spatial levels and ultimately each of them should be matched with the results of the portfolio analysis and the product life cycle analysis of the different tourism and recreation complexes.

At a regional level the general direction of future development is chosen from a set of alternatives. Roughly speaking it is possible to start from two essential types of global strategy:

A. Strategies determining the direction of future development:
* Expansion strategy leading to a larger number of recreationers and/or tourists, but also to greater pressure on scarce resources.
* Consolidation strategy directed only at a qualitative improvement and at maintaining the market share. A modest growth in the use of outdoor recreation facilities is permitted, to benefit the local inhabitants.
* Restrictive strategy. Existing conservational claims for natural areas and the preservation of cultural heritage make it necessary to have a dissuasive policy. This results in a relocation of enterprises (camping sites, for instance) or an extensification of existing infrastructural facilities. Sometimes a demarketing policy is desirable.

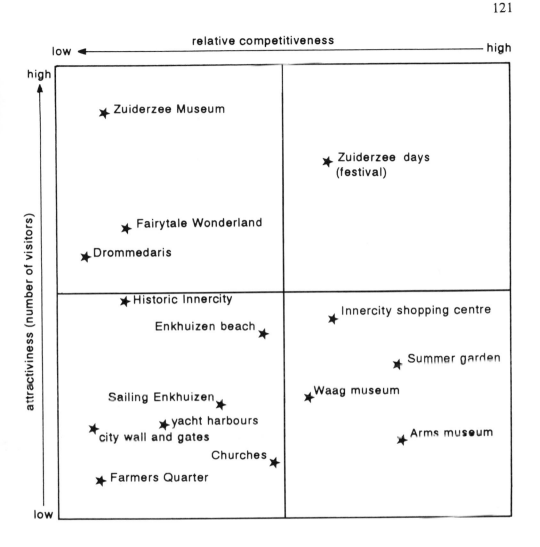

Figure 6.13   PORT FOLIO OF PRIMARY PRODUCT ELEMENTS IN ENKHUIZEN

B. Strategies focusing on the character of the tourism and recreation product itself:
*   Choices for strengthening the particular regional identity of the tourism and recreation product.
*   Choices for consolidating the existing natural resources.

* A strategy for the development of elements important to obtain a position in the international tourism market.

It is important to realize that it is not necessary to choose only one strategy for the whole region in question. The goals and the restrictive conditions may make a spatially differentiated policy desirable. The basic assumption should be the maintenance and improvement of the spatial quality.

6.9 Summary and conclusions

In the 1980s societal processes made it apparent that the existing supply of services and amenities for tourism and recreation no longer met the requirements of the population. It was realized that a more flexible and more market-oriented approach to physical planning was necessary. The most striking characteristic of these societal changes was the phenomenon of differentiation. This ongoing process of differentiation is discussed. Demographic, cultural, technological and socio-economic changes resulted in great differences in 'recreation styles' and, as a consequence, in a variety of tourism and recreation behavior. The supply side of the market has to cope with the very different wishes and expectations of very different user groups. Even the behavior of specific individuals is not constant during a certain time period. A time-budget analysis in West Germany has shown lower participation in activities that take more than two hours. Modern leisure behavior is 'butterfly behavior', an untiring wish for new excitements and challenges.

This process of societal differentiation can only be met by spatially integrating tourism and recreation services and amenities into complexes. The creation of these complexes is embedded in a process of stimulative planning. Within this perspective an outline for a new planning model for tourism and recreation is presented. The renewal is found in the way tourism and recreation complexes have to be identified. The experience of the potential or actual visitor is crucial in determining the relationship between the complexity of the spatial context and the more individual leisure experiences. Interesting insights into the strength and weakness of the tourism and recreation complexes constructed according to the proposed method can be derived by applying techniques used by strategic management. Attempts to remain competitive can be more successful if the proposed model for physical planning is followed, because a market orientation is really present. The application and use of GIS (Geographical Information Systems) can be extremely helpful for this type of planning.

LITERATURE

Andersson, A. and H. de Jong, 1987. Recreatie in een veranderende maatschappij. Deel 1: een literatuurstudie, Mededelingen van de Werkgroep Recreatie, 1, Landbouwuniversiteit, Wageningen.

Ashworth, G.J. and H. Voogd, 1987. Geografische marketing, een bruikbare invalshoek voor onderzoek en planning, Stedebouw en Volkshuisvesting, pp. 85-90.

Ashworth, G. and H. Voogd, 1988. Marketing the city. Concepts, processes and Dutch applications, Town Planning Review, 65-79.

Ashworth, G.J. and M.C. Jansen-Verbeke, 1989. Functional association: the precondition for the use of leisure as a vehicle for the economic revitalisation of inner areas of cities, Pre-Congressbook, vol. 2a, ELRA-Congress Cities for the Future, Stichting Recreatie, Recreatiereeks nr. 6.

Bakker, J.G., H.W.J. Boerwinkel, K. Kerkstra and J.F.B. Philipsen, 1989. Ruimtelijke scenario's voor de functie-afstemming recreatie-natuur. Onderzoeksvoorstel, Werkgroep Recreatie en Vakgroep Ruimtelijke Planvorming, Landbouwuniversiteit Wageningen.

Blaas, H., 1989. Is er nog toekomst voor de landelijke gebieden van Nederland? Planologische Diskussiebijdragen, Deel I, Delft, 107-116.

Boelens, L., 1988. De nieuwe opleving binnen de stedebouw en planologie, een dood spoor? Stedebouw en Volkshuisvesting, 75-81.

Cammen, H. van der, 1988. Environmental planning in the Netherlands in the 21st century. In: L.H. van Wijngaarden-Bakker and J.J.M. van der Meer, Spatial sciences, research in progress, Nederlandse Geografische Studies, 80, Amsterdam, 84-90.

Chapin, F.S., 1974. Human activity patterns in the city, New York, Wiley.

Comijs, J., 1989. Marktgerichte stedelijke planning, Ruimtelijke Verkenningen 1989, Den Haag, 119-127.

Cooper, C., 1989. Tourist product life cycle, in: S.F. Witt and L. Moutinho (eds.), Tourism Marketing and management Handbook, New York, 577-581.

Dietvorst, A.G.J. and E. Wever, 1977. Changes in the pattern of information exchange in the Netherlands, 1967-1974, Tijdschrift voor Econ. en Soc. Geografie, 68(1977)2, 72-82.

Dietvorst, A., 1989. Complexen en netwerken: hun betekenis voor de toeristisch-recreatieve sector, Inaugural address, Agricultural University, Wageningen.

Dietvorst, A. and R.J.P.A. Spee, 1990. Planning en beleid voor de toeristisch-recreatieve sector, Werkgroep Recreatie Landbouwuniversiteit, Wageningen.

Dietvorst, A., W. Poelhekke and J. Roffelsen, 1987. Hoe meer verscheidenheid aan attracties, des te aantrekkelijker de stad, Recreatie en Toerisme (11), 441-445.

Dietvorst, A. and M.C. Jansen-Verbeke, 1988. De binnenstad: kader van een sociaal perpetuum mobile, Nederlandse Geografische Studies, 61, Amsterdam/Nijmegen.

Dijkhuis-Potgieser, H., 1988. Verblijfsrecreatie in Nederland. In: Ruimtelijke Verkenningen 1988, RPD, Den Haag, 48-56.

Engelsdorp Gastelaars, R. van, 1989. Nederland ontwikkelt zich tot een pluriforme samenleving, Recreatie en Toerisme, 361-364.

Floor, H., 1990. Aktiviteitensystemen en bereikbaarheid. In: J.Floor, A.L.J. Goethals en J.C. de Koning, Aktiviteitensystemen en bereikbaarheid, SISWO, Amsterdam.

124

Gans, H., 1986. Vergrijzing en ruimtelijk beleid. Planologische Diskussiebijdragen, Delft, 249-258.

Huppes, T., 1987. Over netwerken en organisaties. Ec.Stat.Ber., 72 (3626), 932-936/940.

Jansen-Verbeke, M.C., 1988. Leisure, recreation and tourism in inner cities: explorative case-studies, Nederlandse Geografische Studies, 58, Amsterdam/Nijmegen.

Jurgens, C.R., 1991. Introduction to GIS-applications in land use planning. Department of Physical Planning and Rural Development, Agricultural University, Wageningen.

Kleefmann, F., 1984. Planning als zoekinstrument, Planologische Verkenningen, 5, VUGA, Den Haag.

Kotler, Ph., 1988. Marketing management. Analysis, Planning, Implementation and Control, New York.

Lengkeek, J., 1989. Recreative qualities of cities: what, where, for whom. Relevance of recent research for revitalisation of urban environment, Pre-Congressbook, vol. 2b, ELRA-Congress Cities for the Future, Stichting Recreatie, Recreatiereeks nr. 6.

Lew, A.A., 1987. A framework of tourist attraction research, Annals of Tourism Research, Vol. 14(4), 553-575.

Nystuen, J.D. and M.F. Dacey, 1961. A graph theory interpretation of nodal regions. Papers and proceedings of the Regional Science Association, 7(1961), 29-42.

Philipsen, J.F.B., 1989. Door privatisering dreigen groepen watersporters uit de boot te vallen, Recreatie en Toerisme, 47-49.

Plog, S.C., 1974. Why destination areas rise and fall in popularity, Cornell HRA Quarterly, 55-58.

Raggers, G.G. and H. Voogd, 1989. Investeren in planvorming, Planologische Diskussie-bijdragen 1989, Delft, 555-564.

Regt, A.L. de, 1989. Kleinschalig landschap in een grootschalig Europa, Ruimtelijke Verkenningen, Den Haag, 12-44.

Rijksplanologische Dienst, 1986. Ruimtelijke Verkenningen 1985-1986, Den Haag.

Smidt, M. de and E. Wever, 1984. A Profile of Dutch Economic Geography, Assen.

Sociaal en Cultureel Planbureau, 1988. Sociaal en Cultureel Rapport 1988, Rijswijk.

Stichting Brug Wageningen (SBW), 1990. Ede in clusters en complexen. Toeristisch-recreatieve structuurvisie voor de gemeente Ede, Wageningen.

Straten, A. van, 1984. Recreatieplanning in onzekerheid? Een probleemverkenning. In: A. van Straten (ed.), Recreatieplanning in onzekerheid, verslag van de studiedag op 15 juni 1984, Werkgroep Recreatie, Landbouwuniversiteit Wageningen.

Thrift, N., 1977. An introduction to time geography, Catmog 13, Norwich.

Tuynte, J. ten and A. Dietvorst, 1988. Musea anders bekeken. Vier Nijmeegse musea bekeken naar uitstralingseffecten en complexvorming. Werkgroep Recreatie en Toerisme, Katholieke Universiteit, Nijmegen.

Voet, J.L.M. van der, 1989. Recreatie en het landelijk gebied, De Landeigenaar, juni, 12-16.

Wijst, T. van der and F. van Poppel, 1985. Economic and social implications of ageing in the Netherlands, NIDI, working papers, nr. 67, Voorburg.

Wissink, G.A., 1987. Nieuwe oriëntaties en werkterreinen voor de planologie, Stedebouw en Volkshuisvesting, 1987(6), 197-205.

Zonneveld, M.M., 1988. Kamperen bij de boer en natuurkamperen in Nederland, Mededelingen van de Werkgroep Recreatie Landbouwuniversiteit, 13, Wageningen.

CHAPTER 7

Recreation management within the multiple use management concept of the United States Forest Service

by
Hubertus J. Mittmann
Regional Landscape Architect
United States Department of Agriculture
Forest Service, Rocky Mountain Region
Colorado, U.S.A.

7.1   Introduction

This paper deals with the recreation management process in the U.S. Forest Service. The concepts discussed were developed by many individuals in research and recreation management and the text quotes much of the Recreation Opportunity Spectrum Users Guide.

National Forests in the United States were established through the Organic Administration Act in 1897 and in 1905 the Forest Service was established in the United States Department of Agriculture. The Forest Service now administers 156 National Forests, 30 Purchase Units, 19 National Grasslands, 13 Land Utilization Projects, 21 Research and Experimental Areas and 39 Other Areas for a net 190.6 million acres.

The Multiple-Use Sustained-Yield Act of 1960 set forth that it is the policy of the Congress that National Forests are established and shall be administered for outdoor recreation, range, timber, watershed, and wildlife and fish purposes. Multiple use means the management of all the various renewable surface resources of the National Forests so that they are utilized in the combination that will best meet the needs of the American people over time. The needs of the American people have changed considerably over time toward a balance between amenity values and product needs. Not only has the recreation use on National Forests increased to 242.3 million visitor days per year, but the recreation needs of the people also have changed from a facility-oriented experience to more dispersed recreation activities and shorter visits.

Since more outdoor recreation occurs on National Forest System lands in the United States than any other single landholding it is imperative that recreation planning and design get proper recognition within the multiple use management process.

126

The following figure portrays the recreation use of the different federal land management agencies in the United States.

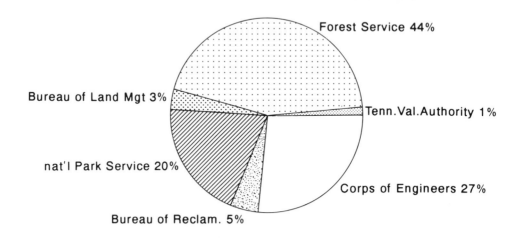

Figure 7.1   RECREATION USE AT FEDERAL AREAS

On National Forest lands many different recreation activities occur, but the majority of use in driving for pleasure and viewing scenery, and camping and picnicking. The percentage of different recreation activities are shown in the following chart.

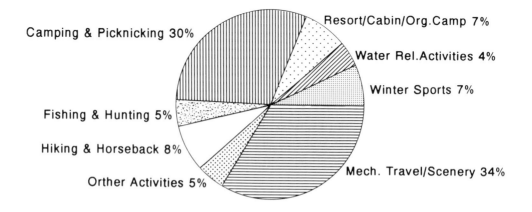

Figure 7.2   RECREATION USE BY ACTIVITY ON NATIONAL FOREST LANDS

Of all these recreation activities 60 percent take place in general forest areas, 20 percent in developed Forest Service facilities, 15 percent in private facilities on National Forest lands and 5 percent in designated Wilderness areas.

A change of recreation activities from facility-oriented recreation experiences to the user dispersed recreation experiences started happening during the past fifteen years. Much research was done during this time to find out how we can best serve the public and provide them with a spectrum of recreation opportunities and the desired recreation experiences.

Based on these changes and needs the USDA Forest Service developed the Recreation Opportunity Spectrum program which will guide the recreation program within this multiple use management agency.

7.2   The Recreation Opportunity Spectrum Program

How important is resource-based outdoor recreation? Where there are finite resources--financial and physical--how do you measure how much support recreation deserves relative to other needs in society? How do you evaluate the benefits which accrue from it?

Evidence from national surveys, Forest Service research, and other data point to leisure as a major element in an individual's personal sense of life satisfaction. A perception of physical and psychological well-being pervades the survey responses regarding recreation. Recreation activity can vary from passive contemplation to strenuous climbing of sheer rock faces. Recreation settings can range from crowded beaches to isolated mountain streams. Regardless of the type of recreation, across the board benefits were cited--as a tonic for physical and psychological weariness and a respite from the day-to-day routine of activities. Psychological increments to the individual include the perception of personal development and self-reliance, communion with nature, a sense of renewal, and relaxation from pressures. Significantly, the priority consideration given to outdoor recreation is consistent with persons on all levels of income, education, and occupational status.

In terms family and community, central elements in people's lives, recreation is a primary link in building and maintaining these necessary social interactions. Family relationships are enhanced when the opportunity for experiencing outdoor recreation settings together result in eased tensions, better communication, and possible long-term behavioral improvements leading to better family cohesion. The shared enjoyments of outdoor recreation cement social relationships between existing and new found friends in the community.

Benefits to society from such school or community-initiated endeavors as participating in ecology projects, can result in increased future demand for the desired physical setting.

Economic benefits resulting from outdoor recreation include improved health and job productivity. Increased tax bases for community services and increased regional income can be brought about by preservation of the resource for recreational activity. Outdoor recreation is a multi-billion dollar industry that provides jobs, and produces goods and services.

The old question arises here on how do you place a dollar value on a sunset? A number of methods have been developed for approximating a dollar valuation of the benefits of recreation. Most have been based on the concept of "willingness to pay". The question is to ascertain what users would pay were the opportunity supplies in a price-elastic market. Since there is no such market, the valuation should include not only what is actually paid but the "consumer's surplus" or worth of the opportunity above the cost.

The basic assumption underlying the ROS is that quality in outdoor recreation is best assured through provision of a diverse set of opportunities. Providing a wide range of settings varying in level of development, access, and other factors, insures that the broadest segment of public will find quality recreational experiences, both now and in the future. Although the notion of quality is relative, a value judgment, the concept of quality can be stated for management decision purposes in this way: quality depends on what experiences the individual is looking for, how much of it is realized, and the degree of satisfaction.

A crucial problem for resource managers, then, is to respond to recreationists' desires for various kinds of appropriate settings managed to produce as many of those experience opportunities as are within the National Forest role. A further challenge is to determine what different practitioners need for satisfying experiences, and if it can be delivered within existing constraints. If a recreation opportunity area is consistently providing satisfactory experiences, the area can be said to be producing quality recreation opportunities, and the users to be receiving full benefit from their experiences. If, on the other hand, there is evidence that inconsistencies exist between what an area offers, what users are led to expect and what managers are trying to provide, the area is producing less than full quality recreational opportunities.

7.3 Recreation opportunity spectrum

While the goal of the recreationist is to obtain satisfying experiences, the goal of the recreation resource manager is one of providing the opportunities for obtaining these experiences. By managing the natural resource setting, and the activities which occur within it, the manager is providing the opportunities for recreation experiences to take place. Therefore, for both the manager and the recreationist, recreation opportunities can be expressed in terms of three principal components: the activities, the setting, and the experience.

For management and conceptual convenience, possible mixes of combinations of activities, settings, and probable experience opportunities have been arranged along a spectrum, or continuum. This continuum is called the Recreation Opportunity Spectrum (ROS) and is divided into six classes. The six classes, or portions along the continuum, and the accompanying class names have been selected and conventionalized because of their descriptiveness and utility in Land and Resource Management Planning and other management applications.

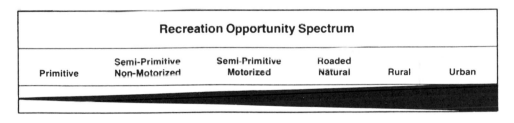

Figure 7.3   RECREATION OPPORTUNITY SPECTRUM (ROS)

Each class is defined in terms of combination of activity, setting, and experience opportunities. Subclasses may be established to reflect local or regional conditions as long as aggregations can be made back to the six major classes for regional or national summaries. An example of a subclass may be a further breakdown of Roaded Natural into subclasses based on paved, oiled, or dirt surfaced roads, which in turn reflects amount of use, or a further breakdown of Primitive based upon aircraft or boat use.

The Recreation Opportunity Spectrum provides a framework for stratifying and defining classes of outdoor recreation opportunity environments. As conceived, the spectrum has application to all lands regardless of ownership or jurisdiction. It's use in the National Forest System will facilitate the consideration, determination and implementation of the recreation management role.

130

The following flow chart outlines the major components of the Recreation Opportunity Spectrum and how it relates to land and resource as well as project planning.

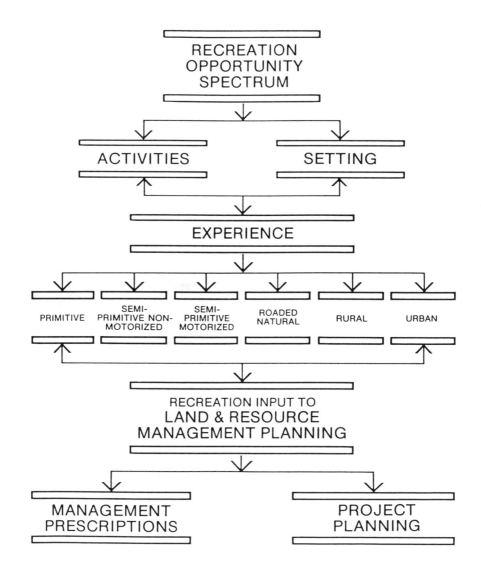

Figure 7.4   MAJOR COMPONENTS OF THE RECREATION OPPORTUNITY SPECTRUM

7.4   Determination of Recreation Opportunity Spectrum Classes

The end product of recreation management is the experience people have. The key to providing most experience opportunities is the setting and how it is managed. The setting criteria have been broken down into three categories to address all influences on the desired experiences:

1. The physical setting incorporates the criteria for remoteness, size and evidence of humans.

    Remoteness refers to the extent to which individuals perceive themselves removed from the sights and sounds of human activity. For a primitive experience the area should be at least three miles removed from motorized user access. On the other end of the scale, to have a rural or urban experience, motorized use would be part of it.

    Location of roads and trails and types of roads and trails are major influencing factors in experience expectations. Size is another important aspect of each experience area. To have a primitive experience the area should be around 5000 acres or larger, To have a semi-primitive non-motorized or semi-primitive motorized experience the area should be around 2500 acres in size. The roaded-natural, rural and urban experience areas do not have any size criteria.

    The third criteria deals with the evidence of human activity for different recreation experiences. On the primitive end it would be a nearly unmodified setting with a strongly structure (buildings, bridges, and the like) oriented setting on the urban end of the scale.

2. The social setting deals with user density. It reflects the amount and type of contact between individuals or groups. It also indicates opportunities for solitude, for interactions with a few selected individuals, or for large group interactions. At the primitive end of the ROS scale very little contact with individuals and groups would be expected.

3. The managerial setting reflects the amount and kind of restrictions placed on people's actions by the administering agency or private land owner which affect recreation opportunities. One would not expect any on-site regimentation when seeking a primitive experience, but regimentation controls in rural or urban settings.

132

The following chart summarizes the different setting criteria for the six recreation opportunity classes.

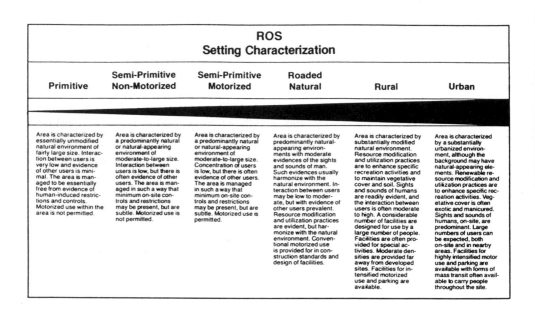

ROS Setting Characterization					
Primitive	Semi-Primitive Non-Motorized	Semi-Primitive Motorized	Roaded Natural	Rural	Urban
Area is characterized by essentially unmodified natural environment of fairly large size. Interaction between users is very low and evidence of other users is minimal. The area is managed to be essentially free from evidence of human-induced restrictions and controls. Motorized use within the area is not permitted.	Area is characterized by a predominantly natural or natural-appearing environment of moderate-to-large size. Interaction between users is low, but there is often evidence of other users. The area is managed in such a way that minimum on-site controls and restrictions may be present, but are subtle. Motorized use is not permitted.	Area is characterized by a predominantly natural or natural-appearing environment of moderate-to-large size. Concentration of users is low, but there is often evidence of other users. The area is managed in such a way that minimum on-site controls and restrictions may be present, but are subtle. Motorized use is permitted.	Area is characterized by predominantly natural-appearing environments with moderate evidences of the sights and sounds of man. Such evidences usually harmonize with the natural environment. Interaction between users may be low to moderate, but with evidence of other users prevalent. Resource modification and utilization practices are evident, but harmonize with the natural environment. Conventional motorized use is provided for in construction standards and design of facilities.	Area is characterized by substantially modified natural environment. Resource modification and utilization practices are to enhance specific recreation activities and to maintain vegetative cover and soil. Sights and sounds of humans are readily evident, and the interaction between users is often moderate to high. A considerable number of facilities are designed for use by a large number of people. Facilities are often provided for special activities. Moderate densities are provided far away from developed sites. Facilities for intensified motorized use and parking are available.	Area is characterized by a substantially urbanized environment, although the background may have natural-appearing elements. Renewable resource modification and utilization practices are to enhance specific recreation activities. Vegetative cover is often exotic and manicured. Sights and sounds of humans, on-site, are predominant. Large numbers of users can be expected, both on-site and in nearby areas. Facilities for highly intensified motor use and parking are available with forms of mass transit often available to carry people throughout the site.

Figure 7.5   CRITERIA FOR THE SIX RECREATION OPPORTUNITY CLASSES

In addition to the settings the different recreation activities which people want to engage in should be considered for the entire recreation opportunity spectrum to assure consistency with the setting requirements. The following activities are shown within the spectrum of recreation opportunities for illustrative purposes. Additional and different types of activities must be considered for specific locations and new ones must be added as they evolve (see figure 7.6).

By combining the setting characteristics and the type of activities people want to engage in we can predict what kind of experiences they will have within the spectrum of recreation opportunities. The following chart indicates the kind of recreation experience people can expect in the different recreation opportunity classes (see figure 7.7).

ROS Activity Characterization					
Primitive	Semi-Primitive Non-Motorized	Semi-Primitive Motorized	Roaded Natural	Rural	Urban

Land Based:

Viewing Scenery
Hiking and Walking
Horseback Riding
Tent Camping
Hunting
Nature Study
Mountain Climbing

Water Based:

Canoeing
Other Watercraft (non-motorized use)
Swimming
Fishing

Snow and Ice Based:

Snowplay
X-Country Skiing/Snowshoeing

Land Based:

Viewing Scenery
Automobile (off-road use)
Motorcycle and Scooter Use
Specialized Landcraft Use
Aircraft Use
Hiking and Walking
Horseback Riding
Camping
Hunting
Nature Study
Mountain Climbing

Water Based:

Boating (powered)
Canoeing
Sailing
Other Boating
Swimming
Diving (skin or scuba)
Fishing

Snow and Ice Based:

Ice and Snowcraft Use
Skiing, Downhill
Snowplay
X-Country Skiing/Snowshoeing

Land Based:

Viewing Scenery
Viewing Activities
Viewing Works of Human-Kind
Automobile (includes off-road use)
Motorcycle and Scooter Use
Specialized Landcraft Use
Train and Bus Touring
Aircraft Use
Aerial Trams and Lifts Use
Hiking and Walking
Bicycling
Horseback Riding
Camping
Picnicking
Resort and Commercial Services Use
Resort Lodging
Recreation Cabin Use
Hunting
Nature Studies
Mountain Climbing
Gathering Forest Products
Interpretive Services

Water Based:

Tour Boat and Ferry Use
Boat (Powered)
Canoeing
Sailing
Other Watercraft Use
Swimming and Waterplay
Diving (skin and scuba)
Waterskiing and Water-Sports
Fishing

Snow and Ice Based:

Ice and Snowcraft Use
Ice Skating
Sledding and Tobogganing
Downhill Skiing
Snowplay
X-Country Skiing/Snowshoeing

Land Based:

Viewing Scenery
Viewing Activities
Viewing Works of Human-Kind
Automobile (includes off-road use)
Motorcycle and Scooter Use
Train and Bus Touring
Aircraft Use
Aerial Trams and Lifts Use
Hiking and Walking
Bicycling
Horseback Riding
Camping
Picnicking
Resort and Commercial Services Use
Resort Lodging
Recreation Cabin Use
Hunting
Nature Studies
Gathering Forest Products
Interpretive Services
Team Sports Participation
Individual Sports Participation
Games and Play Participation

Water Based:

Tour Boat and Ferry Use
Boat (Powered)
Canoeing
Sailing
Other Watercraft Use
Swimming and Waterplay
Diving (skin and scuba)
Waterskiing and Watersports
Fishing

Snow and Ice Based:

Ice and Snowcraft Use
Ice Skating
Sledding and Tobogganing
Downhill Skiing
Snowplay
X-Country Skiing/Snowshoeing

Figure 7.6   ACTIVITIES WITHIN THE SPECTRUM OF RECREATION OPPORTUNITIES

These are the major components of the recreation opportunity spectrum program which lead to the different recreation experiences.

The next step is a determination of where tthe different ROS classes are located on the ground through a detailed inventory.

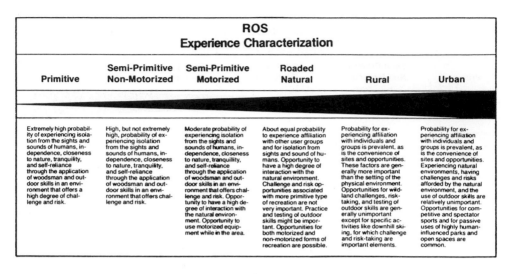

Figure 7.7   RECREATION   EXPRIENCES   IN   THE   RECREATION
OPPORTUNITY CLASSES

7.5   The Inventory Process

The inventory of recreation opportunities should be conducted on a regional scale to assure that all opportunities are considered. The inventory identifies what is actually happening on the land and applies uniformally across all land areas. The inventory is done by analyzing the physical, social and managerial components of the setting and by identifying the remoteness, size, evidence of humans, user density and managerial regimentation which now exist.

The map scale should be chosen to allow an overall view of the planning areas and should be consistent with the mapping scale of the geographic information system program.

Once the ROS classes are mapped on the basis of setting components, the activity opportunities within the ROS classes are identified, and the current capacity of the planning area to provide the different experiences is estimated.

Attractiveness by area and ROS class may also be inventoried if relevant to the analysis of issues and concerns or other management planning needs.

The visual appearance of the land and visual quality objectives for each ROS class are determined through the Visual Management System which is part of the National Forest Landscape Management program.

An overlay system of maps is most effective in deriving at the final delineation of the different recreation opportunities. The following are some of the specific inventories which must be accomplished for the different settings.

7.6   Physical Settings - Remoteness

Delineate all roads, railroads and trails. Distinguish between two levels of roads, "primitive roads", and "better than primitive road" category. The road and trail patterns are mapped. Where densely roaded areas occur only the roads along the periphery of the area need to be mapped and the area itself is identified as densely roaded. Where available, traffic volumes should be recorded. Where air and motorized water travel routes provide the only access to specific areas they must be treated in the same manner as roads.

Using the distance guidelines shown in the following figure 7.8 a remoteness overlay is developed. Lines between ROS classes should reflect topographic differences which screen out sights and sounds of humans. When topographic features act as barriers the remoteness distances may be reduced accordingly.

Remoteness Criteria					
Primitive	Semi-Primitive Non-Motorized	Semi-Primitive Motorized	Roaded Natural	Rural	Urban
An area designated at least 3 miles from all roads, railroads or trails with motorized use	An area designated at least ½-mile but not further than 3 miles from all roads, railroads or trails with motorized use; can include the existence of primitive roads and trails if usually closed to motorized use.	An area designated within ½-mile of primitive roads or trails used by motor vehicles; but not closer than ½-mile from better than primitive roads.	An area designated within ½-mile from better than primitive roads, and railroads.	No distance criteria.	No distance criteria.

Figure 7.8   DISTANCE GUIDELINES USED TO DEVELOP REMOTENESS CRITERIA

The next step in the inventory is to delineate areas which indicate the different degrees of evidence of humans. Based on the criteria outlined in the following figure 7.9, a map overlay is prepared to delineate the areas for all six ROS Classes.

Evidence of Humans Criteria

Primitive	Semi-Primitive Non-Motorized	Semi-Primitive Motorized	Roaded Natural	Rural	Urban
Setting is essentially an unmodified natural environment. Evidence of humans would be unnoticed by an observer wandering through the area.	Natural* setting may have subtle modifications that would be noticed but not draw the attention of an observer wandering through the area.	Natural* setting may have moderately dominant alterations but would not draw the attention of motorized observers on trails and primitive roads within the area.	Natural* setting may have modifications which range from being easily noticed to strongly dominant to observers within the area. However from sensitive** travel routes and use areas these alterations would remain unnoticed or visually subordinate.	Natural* setting is culturally modified to the point that it is dominant to the sensitive** travel route observer. May include pastoral, agricultural, intensively managed wildland resource landscapes, or utility corridors. Pedestrian or other slow moving observers are constantly within view of culturally changed landscape.	Setting is strongly structure dominated. Natural or natural-appearing elements may play an important role but be visually subordinate. Pedestrian and other slow moving observers are constantly within view of artificial enclosure of spaces.
Evidence of trails is acceptable, but should not exceed standard to carry expected use.	Little or no evidence of primitive roads and the motorized use of trails and primitive roads.	Strong evidence of primitive roads and the motorized use of trails and primitive roads.	There is strong evidence of designed roads and/or highways.	There is strong evidence of designed roads and/or highways.	There is strong evidence of designed roads and/or highways and streets.
Structures are extremely rare.	Structures are rare and isolated.	Structures are rare and isolated.	Structures are generally scattered, remaining visually subordinate or unnoticed to the sensitive** travel route observer. Structures may include power lines, micro-wave installations and so on.	Structures are readily apparent and may range from scattered to small dominant clusters including power lines, microwave installations, local ski areas, minor resorts and recreation sites.	Structures and structure complexes are dominant, and may include major resorts and marinas, national and regional ski areas, towns, industrial sites, condominiums or second home developments.

Figure 7.9   EVIDENCE OF HUMANS FOR ALL ROS CLASSES

As a result of completing the remoteness, size, and evidence of humans inventory, the physical setting map can now be completed (see figure 7.10).

Based on the social setting and managerial setting criteria shown in the following two figures (7.10 and 7.11) prepare an overlay which reflects the situation on the ground.

By combining the physical, social, and managerial overlays a map is produced which represents the Recreation Opportunity Spectrum classes on the ground. This map indicates what opportunities are now available to enjoy the different recreation experiences. This is the inventory base map which will be used in the land and resource management planning process.

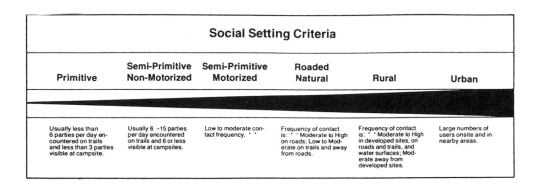

Social Setting Criteria					
Primitive	Semi-Primitive Non-Motorized	Semi-Primitive Motorized	Roaded Natural	Rural	Urban
Usually less than 6 parties per day encountered on trails and less than 3 parties visible at campsite.	Usually 6 –15 parties per day encountered on trails and 6 or less visible at campsites.	Low to moderate contact frequency. ˙ ˙	Frequency of contact is: ˙ ˙ Moderate to High on roads; Low to Moderate on trails and away from roads.	Frequency of contact is: ˙ ˙ Moderate to High in developed sites, on roads and trails, and water surfaces; Moderate away from developed sites.	Large numbers of users onsite and in nearby areas.

Figure 7.10   THE SOCIAL SETTING CRITERIA

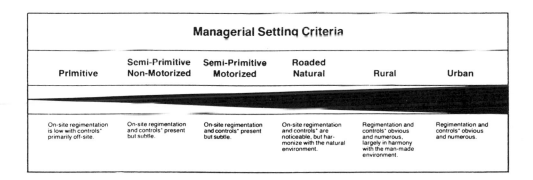

Managerial Setting Criteria					
Primitive	Semi-Primitive Non-Motorized	Semi-Primitive Motorized	Roaded Natural	Rural	Urban
On-site regimentation is low with controls˙ primarily off-site.	On-site regimentation and controls˙ present but subtle.	On-site regimentation and controls˙ present but subtle.	On-site regimentation and controls˙ are noticeable, but harmonize with the natural environment.	Regimentation and controls˙ obvious and numerous, largely in harmony with the man-made environment.	Regimentation and controls˙ obvious and numerous.

Figure 7.11   THE MANAGERIAL SETTING CRITERIA

To give the land manager more information three additional overlays usually are prepared.

1.   Seasonal Use

Land areas which have issues, concerns, and opportunities relating to both summer and winter opportunities may have to be mapped for each season. Activities, setting and experience opportunities may change significantly between the seasons as a result of travel restrictions, accessibility, and apparentness of the evidence of humans criteria.

2.   Attractiveness

This inventory gives the manager a comparison from the attractiveness standpoint. In the decision making process it can be used as a best buy comparison. The attractiveness inventory is based on physical features such as land form diversity, rock form, vegetation types, and water form diversity. This can be the same inventory which determines variety classes in the Visual Management System.

3.   Activity Opportunities

Areas are identified for those recreation activities that require specific settings which are capable of sustaining the impacts the activities put on the land. Examples of these activities are rock climbing, shooting ranges, kayaking, motorcycle or jeep challenge areas and all other activities which require special settings.

7.7   Land and Resource Management Planning

Through the Forest and Rangeland Renewable Resources Planning Act of 1974 and the National Forest Management Act of 1976, the U.S. Forest Service has been directed by Congress to prepare National Forest plans for all the renewable resources on the land and update these plans periodically.

To plan for all the multiple uses so that they are utilized in the combination that will best meet the needs of the American people over time has become a difficult process because needs and desires have changed very rapidly and the importance of amenity values has become very evident. The Recreation Opportunity Spectrum program and the Visual Management System have been developed by the U.S. Forest Service to help plan for these amenity values. Because we have to consider all resources equally in the planning process a specific and complete recreation inventory is required which will be used in combination with all other resource inventories in the development of management plans for all National Forest lands.

The Recreation Opportunity Spectrum serves as the basis for all recreation management plans. Present recreation use and future trends which are predicted through research findings will guide the process of developing recreation management alternatives and the selection of the alternative which best meets the people's needs.

Through the planning process it will be determined what spectrum of recreation opportunities each National Forest will manage for and where these opportunities are located. Once this determination has been made, all other resource management activities will be planned and designed to meet the setting requirements for each ROS class.

By knowing where the different ROS class are managed for, the agency can now prepare visitor maps to guide recreation visitors to the areas which offer the opportunities for the experiences they are seeking.

For successful program implementation, a monitoring system must be developed and applied to assure that the social setting criteria are met and that management plans and designs for other resources meet the setting criteria for the different ROS classes.

This program is now being applied by the U.S. Forest Service in managing 190.6 million areas of land under the multiple use management principle. The concept has proven to be effective and workable and will guide future recreation management.

Information on the ROS program can be obtained from USDA Forest Service offices throughout the United States.

REFERENCES

U.S. Department of Agriculture, Forest Service, 1988. Land Areas of the National Forest System, FS-383.

U.S. Department of Agriculture, Forest Service, 1982. ROS User Guide. August 1982.

U.S. Department of Agriculture, Forest Service, 1986. ROS Book. The book is a combination of the ROS User Guide, Planning Guidance, Research Findings, and Sources of Literature.

U.S. Department of Agriculture, Forest Service, 1987. Project Planning, ROS User's Guide. Chapter 60 (Advance Copy).

U.S. Department of Agriculture, Forest Service, 1983. The Principle Laws Relating to Forest Service Activities, Agriculture Handbook No. 453.

CHAPTER 8

New developments and concepts in tourism and recreation planning in Switzerland

by

Janos Jacsman, René Ch. Schilter and Willy A. Schmid
Institute for National, Regional and Local Planning ETH
(Swiss Federal Institute of Technology) Zurich, Switzerland

8.1   Preamble

In October 1990, in Denver, Colorado, USA, the ISOMUL (International Study Group on Multiple Use of Lands), in conjunction with the American Council of Educators in Landscape Architecture, conducted an international symposium on new developments and concepts in tourism and recreation. The aim of the conference was to present, to compare and to evaluate the methods, ideas, approaches and concepts which have been developed over the past few years in the USA and the countries of Western Europe. Recommendations for research and practice were a part of the findings of this symposium, and the Institute for Local, Regional and National Planning of the ETH Zurich was invited to supply a Swiss contribution to the program. Included in this presentation of tourism and leisure planning in Switzerland were:

a) Specific skeleton conditions
b) New developments in research and planning
c) Successful examples.

This chapter reports on the Swiss contribution to the symposium.

8.2   Skeleton conditions of tourism and recreation planning in Switzerland

8.2.1   The cultural-historic background

Switzerland is a country with a long touristic tradition. Tourism to locations with healing spa waters has existed since the time of the Romans. However, tourism in its modern sense only began in Switzerland in the early 19th century when its mountains and lakes were discovered to be a tourist attraction. Even so, visitation was limited to the summer. Wealthy guests--nobles, real-estate owners, officers' families and other representatives of the upper social class--came from Great Britain and later from other European countries and climbed the mountains or lingered by the lakes. The first hotels were built. A few mountain villages set about becoming tourist centers. The general growth of prosperity, the introduction of legally prescribed vacations and the development of improved modes of transportation (railway, steam-ships) towards the turn of the century promoted the new tourism and strengthened its commercialization. Numerous hotels with the eminent "palace" architecture began to spring up. Mountain railways, and sports and recreational facilities were constructed. Many tourist resorts attained international acclaim. The First World War interrupted the "Belle Époque" of tourism in Europe, and in Switzerland. In the subsequent confusion and revolutions, many kingdoms - and with them the aristocracy which lent the tone to tourism - came to an end. Tourism worldwide experienced its first economic crisis. The attempt to revive tourism in the twenties and thirties brought only partial success.

A new age of tourism commenced after World War Two and has continued until today with some smaller, recession-conditioned setbacks. The most significant features of the new upswing are:

a) Democratization of tourism
b) Supplementation of hotellery through para hotellery
c) Expansion of winter tourism
d) Spread of weekend and excursion tourism.

The economic growth after the Second World War, the increase in the working population's leisure time, the expansion of private motoring and the diminishing environmental quality of residential areas (as a consequence of urbanization and industrialization) all contributed to a wide social dissemination of the touristic demand, and led to mass tourism. The expansion of air traffic led to its internationalization. Growing demands en masse could no longer be handled by the hotel industry alone. Chalets, vacation apartments, and group accommodation facilities were built and camping sites were constructed. The change in structure had its effect on the seasonal spread of the demand. Summer tourism in Switzerland had a rival in the development of attractive and more favorable bathing resorts in the Mediterranean region, and Swiss tourism began to stagnate. On the other hand, winter tourism was to experience a rapid growth. Following the downhill skiing boom came the wave of cross-country skiing. Even by the end of the seventies half the total tourist turnover was achieved during the winter season, although the summer season even now is more important from the aspect of quantity (table 8.1).

TABLE 8.1   CATEGORIES OF GUESTS

Overnight accommodations in hotellery, spa-hotellery and para-hotellery, in billions								
Origin of guests	1975	1980	1984	1985	Winter 83/84	Summer 1984	Winter 84/85	Summer 1985
Switzerland	36.0	39.3	39.5	39.5	17.6	21.9	17.2	22.5
Foreign countries (Total)	32.2	36.0	35.5	35.2	14.9	19.9	15.6	19.6
Fed.Rep.Germany	12.8	16.4	14.4	14.5	7.3	7.0	7.7	6.9
USA	2.1	1.9	3.3	3.6	0.8	2.5	0.9	2.6
Netherlands	3.1	4.0	3.0	2.8	1.0	1.9	1.0	1.7
Great Britain	1.8	1.9	2.7	2.7	1.1	1.7	1.1	1.6
France	3.3	2.8	2.5	2.5	1.3	1.2	1.3	1.2
Asia	1.0	1.3	1.9	1.8	0.5	1.3	0.6	1.2
Belgium	3.1	2.8	1.8	1.8	0.8	1.0	0.8	1.0
Italy	1.2	1.0	1.2	1.2	0.5	0.7	0.5	0.7
Scandinavia	0.7	0.6	0.7	0.7	0.3	0.4	0.3	0.4
Austria	0.5	0.5	0.5	0.5	0.2	0.3	0.2	0.3
Latin America	0.5	0.5	0.4	0.5	0.1	0.3	0.2	0.3
Other countries	2.1	2.3	2.6	2.6	1.0	1.6	1.0	1.7
Overall Total	68.2	75.3	74.5	74.7	32.5	41.8	32.8	42.1

Winter: November-April; Summer: May-October
Source: Schweizer Tourismus in Zahlen, 1989

A special feature of postwar tourism is the spread of weekend and excursion tourism. While weekend tourism as a rule depends on overnight accommodation facilities in a specific region (and is therefore only preferred by a certain section of the population), excursion tourism without overnight stay is open to the total population. In the forties, nearby recreation--as excursion tourism is usually called-- was still mainly limited time wise to the warm seasons, and location wise to the cities and their agglomerations.

For the past thirty years all classes of the population have participated in nearby leisure, whereby excursions have spread over the whole year and location wise extend into the wider vicinity from the place of residence (figure 8.1).

144

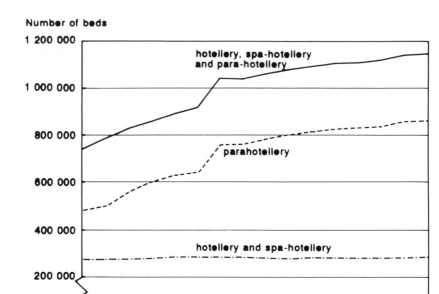

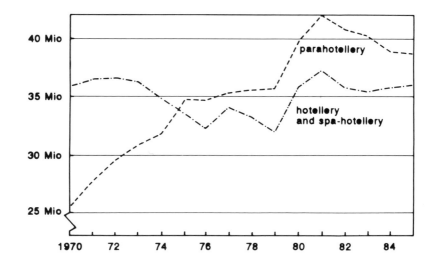

Figure 8.1   OVERNIGHT ACCOMMODATIONS IN SWITZERLAND (top)
AND OVERNIGHT ACCOMMODATIONS IN HOTELLERY, SPA-
HOTELLERY AND PARA-HOTELLERY (bottom)

Source: Schweizer Tourismus in Zahlen, 1981 and 1989

8.2.2 The economic importance

Tourism represents one of the major sectors of Switzerland's economy. In 1988 the total income from tourism amounted to SFr. 16.9 billion. This represents 6% of the Swiss gross national product. Especially important for Switzerland is the income from foreign guests, which has the same effect on the Swiss balance of payments as the export of goods. In this light, 10% of the income from exports can be ascribed to tourism, and for years this has represented the third largest "export sector". The income from foreign guests--at SFr. 10.3 billion--is counter-balanced by Swiss tourists spending SFr. 8.8 billion abroad. The result is a surplus in the balance of payments of SFr. 1.5 billion.

Contrary to vacation and weekend tourism the economic importance of excursion traffic in the earlier years was considered unimportant. Surveys have shown that during the eighties SFr. 2 billion were spent on excursions in Switzerland (Schmid-hauser, 1983). This means that about 30% of the income from domestic tourism is due to day excursions (table 8.2).

With about 220,000 employees (1988), tourism is one of Switzerland's major employers. About 7.4% of the total working population is directly employed in this sector. By including employees who indirectly depend on tourism this proportion could well amount to 12%. In the tourist industry about 180,000 jobs are offered by the catering industry. Of these, two thirds are ascribed to immediate touristic requirements.

With this substantial role in the employment market, tourism fulfills an important regional function in Switzerland. Most of the jobs in tourism are in mountainous areas where otherwise only limited agricultural and forestry jobs would be available. Through its close interlinking with the rest of the economy, tourism simultaneously promotes other economic activities, especially the building industry. The relatively large and differentiated scope of additional work and earning capacity counteracts the acute rural exodus in mountainous regions. Tourism alone cannot solve economic problems in mountainous areas, but without it there is no alternative (Elsasser et al., 1982).

146

TABLE 8.2   THE ECONOMIC SIGNIFICANCE OF TOURISM

Total income from tourism	1978 billion Fr.	1983 billion Fr.	1987 billion Fr.	1988 billion Fr.
domestic guests	3.9	5.9	6.6	6.6
foreign guests	5.6	8.6	10.0	10.3
Total	9.5	14.5	16.6	16.9
Share on the Swiss gross national product (GNP)	6.0%	6.8%	6.8%	6.1%

Income from exports of economic sectors	1978 billion Fr.	1983 billion Fr.	1987 billion Fr.	1988 billion Fr.
1. Mechanical industry	18.3	23.9	31.0	32.5
2. Chemical industry	8.4	11.5	14.6	15.9
3. *Tourism*	*5.6*	*8.6*	*10.0*	*10.3*
4. Textile industry national product (GNP)	3.0	3.9	4.3	4.4
5. Watch industry	3.4	3.4	4.3	5.1

Income/expenditures	Balance from tourism (in billion Fr.)							
	income from foreign guests				expenditures by Swiss tourists abroad			
	1978	1983	1987	1988	1978	1983	1987	1988
overnight tourism	3.1	4.5	5.1	5.2	-	-	-	-
study and spa-tourism	0.5	1.0	1.2	1.3	2.2	3.8	5.2	5.9
day excursion and transit traffic	0.8	1.1	1.7	1.7	0.8	1.0	1.3	1.5
international passenger transport	1.2	2.0	2.0	2.1	0.7	1.1	1.3	1.4
Total	5.6	8.6	10.0	10.3	3.7	5.9	7.8	8.8

Source: Schweizer Tourismus in Zahlen 1989

Share of employees

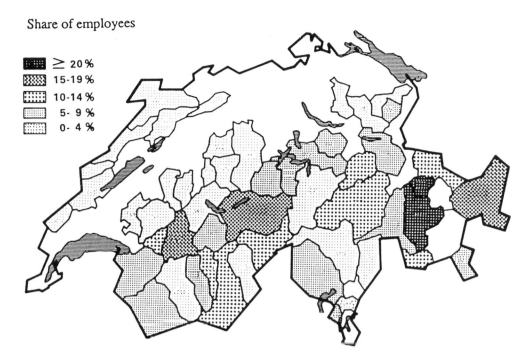

$\geq 20\%$
15-19%
10-14%
5- 9%
0- 4%

Figure 8.2   SHARE OF EMPLOYEES IN HOTELLERY FOR THE IHG-
REGIONS, 1981

Source: Elsasser and Leibundgut, 1984

8.2.3 Legal foundation

At the Federal level Switzerland does not have any special legal foundation for
tourism and recreation. There are, however, numerous legal regulations which in
their application serve indirectly the promotion or realization of touristic and
recreation planning. They only come into use wherever the Cantons or the
Communities are not active. In most Cantons for which tourism is of importance
there is a nominal touristic legislation that has been largely built up over the past 20
years.

From the functional, touristic-relevant Federal laws, the legislation on working
times, vacations and national holidays should be mentioned first. Although
Switzerland is counted among the most liberal of the industrial countries, because of
the legislation on legal contracts, the 42-hour working week with 5 working days
and the 4-week minimum vacation have been accepted. Secondly, the most

important functional legal basis in the provision of recreation and tourism are the urban and regional planning law, the forestry law, the law on the preservation of nature and countryside, the water protection law, and the environmental protection law. They enforce the protection of natural landscapes and the environmental planning of tourist settlements, and therefore the safeguarding and development of attractive recreational areas that are the basic assets of tourism. The legislation applying to the promotion of touristic facilities and the granting of concessions for tourist transport installations must also be mentioned.

A third group of legal foundations promotes the freedom of movement and activities by tourists: the arrival in the country, accommodation in the resort and easy traffic across Swiss borders.

Finally, there are the national regulations applying to the operation of touristic installations, the qualifications of the personnel and the supervision and financing of hotel and resort credits.

8.2.4   The actual problem situation

The strong upswing of tourism in the postwar period effected a massive expansion of the entire touristic offer. However, because of the marked seasonality of tourism in Switzerland, sport and recreational facilities intended for peak periods are utilized only during the few summer and winter months. The over-dimensioned facilities led to high investment costs and correspondingly high operational costs. This meant that the finances of many communities, spas, and transport companies were overstrained. Even in peak months climatic influences can lead to enormous declines in the demand. In order to counteract these declines additional facilities (indoor sports halls in summer and snow cannons in winter) have been constructed that in turn require new investments and more operating costs (table 8.3).

TABLE 8.3   SEASONAL OCCUPATION OF BEDS IN THE HOTELLERY

year	Occupation of existing beds in %			Occupation of available beds in %		
	Summer	Winter	year	Summer	Winter	year
1978	36	27	32	42	36	39
1983	39	29	34	45	37	41
1987	38	29	34	44	37	41
1988	38	29	34	44	37	41

Source: Schweizer Tourismus in Zahlen, 1989

Economic difficulties are intensified because a continued qualitatively good guest service (needing an intensive personnel operation) is demanded of the tourism industry. This is further aggravated by the competition on the employment market by more attractive sectors. The expansion of the touristic offer (estates, transport and recreational installations) has been carried out at the expense of the natural countryside, which in many places has caused considerable strain on the environment. Traffic, with its noise and air-polluting emissions, has caused environmental damage--especially during weekends. The leisure activities in ecologically sensitive areas (the habitat of endangered flora and fauna) can be an equally heavy impact. Especially sensitive to the touristic attack is the Alpine landscape because of its extreme conditions (steeply sloping, moving topography, limited soil substance, shorter vegetation time, and so on). This constant ecological burdening can have the consequence that many life-essential environmental functions might no longer be provided for local inhabitants and tourists. The tendency for natural catastrophes such as avalanches, flooding, and landslides is increasing, while the yield from the soil (agricultural and forestry) is diminishing. Even many of the recuperative effects of the countryside--such as the closeness to nature, quietude and healthy atmosphere--are in doubt (figure 8.3).

A third complexity relates to the behavior of the local population towards touristic development. Although tourism depends on the care and maintenance of natural agrarian countryside, the mountain economy is overstrained by building on life-essential valley soils which are ideally suited for agricultural purposes. Because of tourism-conditioned price increases in building land and real estate, local inhabitants find reasonably priced apartments hard to find, not to mention the purchase of land for their own use. This leads to social and socio-political tensions, increasing the need for relocation of job and residence. These social and socio-political problems are aggravated by the sale of real estate to outside and foreign persons interested in second residences, and the buying up of Swiss land by foreigners. Also to be remembered are the risks of infiltration of the native culture by confrontation with the tourists' lifestyle and conduct.

8.3   New developments in research and planning

8.3.1   The Swiss tourism concept

At the end of August 1973, the Federal Council (as the Swiss government is called) appointed a consulting commission for vacational traffic to coordinate the tourismo-political affairs at national level. The first step of the commission was the devising of an overall Swiss tourism concept. The results were presented in 1979.

NUMBER

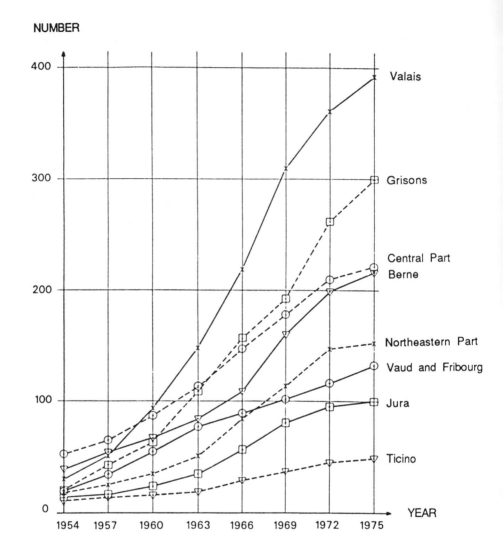

Figure 8.3   NUMBER OF TRANSPORT INSTALLATIONS FOR TOURISM
             IN   THE   DIFFERENT   RESORTS   OF   SWITZERLAND
             DEVELOPMENT BETWEEN 1954 AND 1975

Source: Weiss, 1981

The Swiss tourism concept provided the basis for a future tourism policy. Within
this framework the concept was to assist the federal authorities in coordinating the
various departments dealing with the most important tourismo-political federal tasks
and, via governmental instrument, to help control and influence tourist development
in Switzerland. For other participants in tourism this concept was to be of assistance

when making a decision. In Swiss practice, the concept summarized the aims and measures into a unified statement.

The Swiss tourism concept provides answers to the following questions: (Consulting Commission, 1979):
a)	What important targets should be aimed for in tourism policy?
b)	What strategies should the exponents of the tourism-policy look for when making decisions?
c)	What tourismo-political measures must the Federation take?

The objectives hinge on a simple basic model in tourism policy consisting of three targeted groups which are independent of each other: society, economy and environment (figure 8.4). The whole objective of the tourism policy is to satisfy people's tourist needs by economical use of resources within the framework of an intact environment. The whole objective was carried out by a hierarchic diversion of aims based on the reaction of purpose to means. The result is a multi-layer target system.

The strategies are the connecting links between aims and practical measures. They are in the tension field of society, economy and environment, relate to one or more aims, and are directed to the immediately affected authorities and private persons. The strategies are categorized as:
- idealistic strategies
- institutional strategies
- growth strategies
- marketing strategies
- exploitation strategies
- relief strategies.

The idealistic strategies concern the touristic chances of encounter which should be made use of and the cultural character of the resorts promoted. The institutional strategies concern the increasing inclusion of the local population in a copartnership (participation) and protection of the visitor as a consumer. With the growth strategies primarily a qualitative development of tourism is postulated. The marketing strategies demand an economically efficient advertising policy that takes local and regional interests into consideration, and an advertising policy based on improved market research. The exploitation strategies show how future long-term development of social and economical interests within the given environmental-conditioned limits can be planned and steered. The relief strategies lessen the load on the environment through touristic use, including, for example traffic emissions and changes in the appearance of communities and countryside.

152

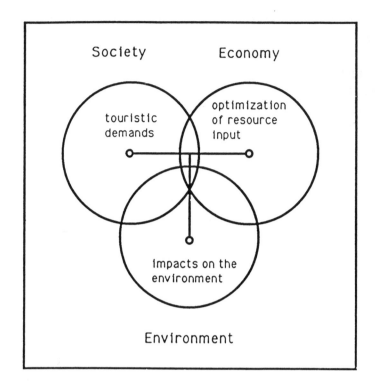

Figure 8.4   BASIC MODEL FOR THE SWISS TOURISM CONCEPT

Source: Beratende Kommission, 1979

The proposed tourismo-political measures are addressed to the exponents of touristic policy at the Federation: parliament, government and administration. A distinction is made between immediate measures and consequent midterm measures. Within the first-mentioned group it is important to firstly accomplish the following:

- Declaration that the tourism concept to be binding;
- Request to work out additional concepts at Cantonal level;
- Strengthening of the organs responsible for tourism policy.

The responsible authorities must be pledged to a restrained concession practice for touristic access facilities and a balancing of the tourism concept within the aims of the zoning policy regulations. The subsequent measures are mainly to safeguard and improve: the existing permanent measures in promoting facilities, national advertising, cooperation among para-state tourism organizations, tourist development aid, touristic research, planning and statistics. New measures are required for consumer protection, social policy and the environment. Additional measures are needed for the economical advancement of tourism. The list of necessary measures is concluded with the demand for checking touristic legislation and for the devising of basic tourist legislation.

In 1981 the Swiss tourism concept was declared binding by the Federal Council. With this move unity in the concept for the Swiss tourism policy was established for the first time. "This was possible because to a certain extent it was a mere target concept, without wieldy measures" (Müller, 1986). In retrospect it must be said that through lack of legislation the effects of the Swiss Tourism concept, especially on regional and local tourism development, have been somewhat modest. It must be clarified whether the tourism concept of 1979 should be completely revised.

8.3.2   Results of the MAB Research Program

The MAB international research program (Man and Biosphere) was launched by UNESCO in 1971 and approved in 1972 by the UN Environmental Conference in Stockholm. Switzerland participated in this program with the project "Socio-economic Development and Ecological Impact Capacity in Mountainous Regions". The project was undertaken between 1979 and 1985 in four test areas (Grindelwald, Aletsch, Pays-d'Enhaut and Davos) within a national research program. The object of the study was described as follows: "On the basis of scientifically proven findings about the actual impact and the tolerable impact capacity of the various ecological systems, firm measures are to be devised for the ecological use of mountainous regions. The base for creating new ecosystems is to be made available and adjusted to changed social conditions now ecologically unacceptable." (MAB 1976). The project was carried out in three stages:

1. Investigating the functions of regional socio-ecosystems.
2. Determining the impact capacity of such systems ecologically and socio-economically.
3. Examining and determining the control mechanisms, taking impact capacity limits, the local population's and the tourists' demands into consideration.

The study took a regional ecosystems approach, the content of which was an optimization of a man versus environment system on the use of land. Open space was restricted to the four areas already mentioned (figure 8.5).

From the title of the project it is clear that tourism represents one of the most crucial problems. A main question of the study was indeed: "How can tourism contribute so that the mountainous regions are able to remain living, economical and recreational areas, and how does it endanger these aims" (Müller, 1986). The analyses were based on the local mountain socio-ecosystems with the following

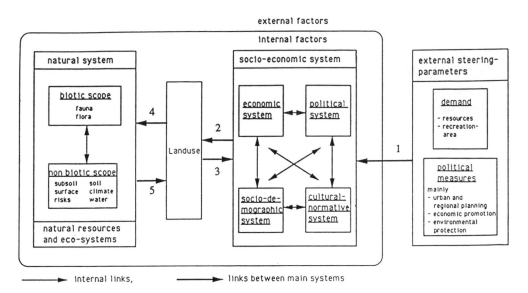

external factors

internal factors

1 External factors and their effects on the socio-economic conditions
2 Socio-economic development and the induced changes of landuse
3 Feedbacks of changes in landuse on population, economy and society
4 Landuse systems and reactions from the external systems
5 Feedbacks of changes of natural systems on the human living conditions

Figure 8.5   SIMPLIFIED SCHEMATIC DESCRIPTION OF A REGIONAL ECOSYSTEM

Source: Messerli and Messerli, 1978.

13 components: landscape, air, water, infrastructure, local populations, tourists, catering, parahotellery, cableways, agriculture, trade, services and community politics.

Analyses have shown that tourism has relation to nearly all these components. In order to record the diverse and complex inter-dependent actions as a whole, and to be able to evaluate them, a "tourist network matrix" was developed. Firstly, it distinguishes between effects from within and from outside, whereby effects caused by a component from within and effects by a component from outside are considered to be tolerable influences.

The weighing up of these leads to the following categories:

- the active components (effects from within dominate those from outside)
- the critical components (considerable effects from within and from outside)
- the passive components (effects from outside dominate those from within)
- the still components (no effects from within or from outside).

The affinities can be positive or negative. The positive ones are called benefit relations; the negative ones damage relations. The function of the active and passive benefit relations and the active and passive damage relations was registered in the damages and benefits relation matrix (figure 8.6).

Four new categories of system components can be sub-categorized as a result of the combination of the benefit and damage comparison:
- "Stars": They produce mainly benefits and are at the same time favored by most effects from outside.
- "Cash cows": They produce mainly benefits, but they are affected more by damage than favored by benefits.
- "Dogs": They mostly cause only damage and are affected throughout by damage.
- "Parasites": They effect greater damage than benefits, but are favored by benefits rather than affected by damage.

The natural elements and system components: landscape, air and water belong to the "cash cows" category. Their function as benefit producers is very strongly marked. At the same time they are affected by distinct damage.

Both social components--local populations and tourists--are the "stars" since they produce more benefits than damage, and because they are affected by more beneficial than damaging effects. At the same time they are the ones who profit from the system because they can leave the system if conditions are not to their liking.

The components' infrastructure is found between "Dogs" and "Cash Cows". Its passive damaging effects are significantly greater than the beneficial effects. Positive and negative effects are held in balance. The economical components throughout prove to be the beneficiaries. Agriculture and catering in this respect are the "stars", while parahotellery is classed under "parasites". As to the others (catering, cableways, trade and services) a clear cataloguing is not possible.

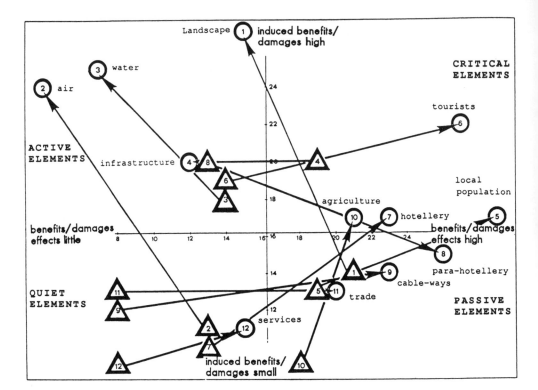

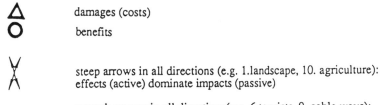

△ damages (costs)
○ benefits

X steep arrows in all directions (e.g. 1.landscape, 10. agriculture):
   effects (active) dominate impacts (passive)

⋈ smooth arrows in all directions(e.g. 6.tourists, 9. cable ways):
   impacts (passive) dominate effects (active)

Figure 8.6   COMPARISON OF DAMAGES AND BENEFITS FOR A
              LOCAL MOUNTAIN-SOCIO-ECOSYSTEM

Source: Müller, 1987

The component community politics is among the "cash cows".

The comprehensive damage/benefits analyses in the tourist network matrix enable a choice between four feasible basic strategies for the control mechanism:

1. Support beneficial effect produced from within.
2. Reduce damage effect caused from within.
3. Support favorable influences from outside.
4. Reduce loads caused from outside.

It is now a matter of finding strategies and measures which greatly improve the damage/benefit equation. To wait for a self-regulation by negative feedback is too risky. The general thrust of the strategies must be oriented toward reducing the speed and intensity of development" (Müller, 1986).

In actual fact the study proposed system-enduring strategies and measures for the environment, society, economy and politics. In the environment it is a matter of reducing the existing impact on natural elements (landscape, air, water) and limiting early enough new damaging effects. For society, the impact on the local populations is to be reduced and their benefits increased. The damage caused by tourists is to be reduced. In the economy it is mainly a question of weighing up the two active strategies of benefit increase and damage reduction. Qualitative instead of quantitative growth is the watchword. Politics requires the creation and use of instruments which will positively influence the damage/benefits equation. For this the benefits will be increased and/or the damage reduced.

The importance of the Swiss MAB study cannot only be assessed for having devised strategies for dealing with a future development of the mountainous regions as living, economical, and recreational areas. Over and above this, the study offers responsible authorities the possibility of checking the Swiss tourism concept on the basis of the latest research results.

8.3.3 Enquiry into the impact through recreation on landscape

In the mid eighties a study on the assumed impact on forests by recreationists was carried out (Jacsman, 1990) at the Institute for National, Regional and Local Planning of the ETH in Zurich. The aim of the enquiry was to quantify the impact on Swiss forests through recreation. The key points were:

a) the simultaneous maximum impact in persons/ha
b) the annual impact in visitor-hours/ha.

By forest leisure a special situative form of recreation in the open countryside was understood, and characterized by the activities of hiking, walking, playing games, lying down, picnicking, and experiencing nature and countryside. Not included were recreational activities in water and snow. The reference space for the survey was the national total surface of Switzerland, which, for this purpose, was split into 210 reference regions. From the point of view of time three impact conditions were taken into account: the condition in 1980 (condition "Yesterday") and two potential impact conditions in the near future. One is based on total motorization of all households and the completion of the national highway network (Autobahnen) in Switzerland (condition "Tomorrow IT"). The other is reckoning with a changeover to public transport by all recreationists (condition "Tomorrow PT").

The estimate of the impact was done by means of an information model of the open space recreation which existed from three part models. The part models of recreational demand determined the demand by the native population, along with the vacation and weekend guest, for nearby natural recreation within the residential district or region of stay, as well as the demand by the native population for nature-near recreation outside the residential district (excursion without overnight stay). The part model of visitor distribution estimated the spread of leisure demand in the targeted regions, and the emphasis lay on the distribution of day tourists. In the part model the impact on open space was the estimated part impact combined, and the quantifying of the sought targeted dimensions. All three part models use a statistic estimated procedure, known as a balance model which makes use of the theory of the stochastic linear systems. A balance model contains no nomologic statements with empirical content. It works with inaccurate information, and input and output variables can be defined according to problems.

In the description of terms, the impact on forests through recreational use was determined from the impact on landscape, consisting of agriculturally-utilized areas, conservation areas and the forests themselves. It can be deduced that primary output data of the models refer to the general recreation in the open countryside. Figure 8.7 shows the maximum simultaneous impact on open space through recreational use in 1980 (condition "Yesterday"). Around 1980 the maximum simultaneous impact on open space in Switzerland amounted on average to about 0.18 to 0.20 visitors per hectare, the annual impact to about 36 to 38 visitor-hours per hectare. The problem regions were likely to be the urban and especially the metropolitan regions. In the latter, a medium maximum simultaneous impact of 1.50 to 2.25 persons per hectare and a medium annual impact of 530 to 690 visitor hours per hectare was shown. For the country regions, including undeveloped mountain regions, there was a result of 0.14 to 0.16 visitors per hectare, or a medium annual impact of 21 to 23 visitor hours per hectare. In the internationally famous tourist regions the maximum simultaneous impact amounted to about 0.15 to 0.30 visitors per hectare and an annual impact of about 50 to 85 visitor hours per hectare. From this it follows that only the general countryside open space recreation in urban

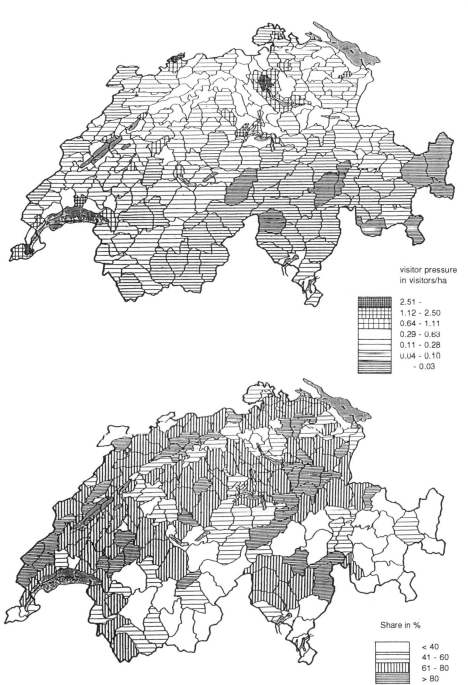

visitor pressure
in visitors/ha

2.51 -
1.12 - 2.50
0.64 - 1.11
0.29 - 0.63
0.11 - 0.28
0.04 - 0.10
 - 0.03

Share in %

< 40
41 - 60
61 - 80
> 80

Figure 8.7   MAXIMUM SIMULTANEOUS IMPACT ON OPEN LANDSCAPE
(top) AND SHARE OF DAY EXCURSIONS ON THESE IMPACTS
(bottom); CONDITION "YESTERDAY" (1980)

160

residential regions is problematic. Neither the daily excursion traffic nor actual tourism contributes in Switzerland to this critical regional impact in the targeted regions.

With total motorization of households the demand for countryside leisure outside residential areas increases. In consequence of the completion of the national highway network, motorized excursion traffic was displaced to the newly-accessible and attractive regions of the Jura, the Lower Alps and the Alps. With the change-over to public transport the demand for countryside leisure outside the residential areas would reduce, and would increase inside the residential areas. The inter-regional recreational traffic would concentrate on the attractive leisure areas well served by public transport. In both cases excursion traffic only would be affected. As seen in table 8.4, the anticipated mutations do not suffice to significantly change the existing impact situation on open space leisure in Switzerland. In particular they would not reduce the critical impacts on urban regions through recreational use.

TABLE 8.4   ANNUAL IMPACT ON OPEN SPACE ACCORDING TO TYPES OF REGION

Region		Visitor-hours/ha according to conditions		
		"Yesterday"	"Tomorrow IT"	"Tomorrow PT"
CR	Medium value	613	610	646
	Distribution	79	80	83
MR	Medium value	205	210	219
	Distribution	16	17	18
PR	Medium value	88	89	89
	Distribution	4	5	5
SR	Medium value	37	39	35
	Distribution	1	1	1
RR	Medium value	22	25	18
	Distribution	1	1	1

CR =   City regions
MR =   Medium-sized urban regions
PR =   Polycentric regions
SR =   Small-sized urban regions
RR =   Rural regions

The proof of stable impact structures of general countryside open space recreation is one of the most interesting results of the study. This stability hinges on the existing settlement structure and Switzerland's far reaching traffic network. The natural recreational suitability in the selection of targeted regions is not a main factor because there is an extensive offer of attractive leisure time locations. Moreover, it can be proved that excursion traffic only negligibly influences the general countryside open space leisure of vacational and weekend guests in a few Lower Alp regions, as figure 8.8 shows.

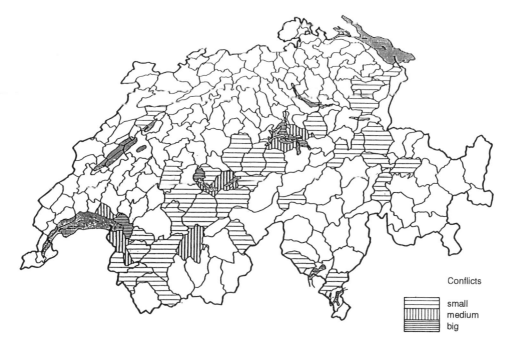

Figure 8.8   POSSIBLE CONFLICTS BETWEEN DAY EXCURSIONISTS AND GUESTS (Condition "Tomorrow IT")

8.3.4   Enquiries into the socio-cultural aspects of tourism

In February 1990 at the Institute for Local, Regional and National Planning ETH, Zurich, a research project entitled "Cultural Diversity, Tourism and Identity in Mountainous Regions" was concluded. The project was carried out within the national research program "Cultural Diversity and National Identity" initiated in 1985. The aim of the project was to show "how a touristic development should be planned if it is to contribute to the promotion, or maintenance, of the cultural diversity and identity" (Bücheler and Elsasser, 1987). The contents were limited to vacational tourism. As a test area for the analytical case studies, Val Calanca, Averstal and the Community of Breil/Brigels were selected. These resorts are all in the Canton of Grisons which is socio-culturally diversely structured. Italian dialect is spoken in Val Calanca; German-speaking Walser inhabitants live in Averstal; and the Community of Breil/Brigels has a Rhetoromansh-speaking population. Both valleys have only small resorts with "soft" tourism. Breil/Brigels is a medium-sized resort with heavy touristic building construction and it contains a skiing area with ski-lift facilities.

The enquiries were based on three working hypotheses (Bücheler, 1990a):

1.   Tourism starts up a culture change in the mountainous area, in which the cultural diversity of the mountainous region is partially lost.
2.   Immaterial influences of tourism have a great effect on places and regions that are existentially (materially) dependent on tourism.
3.   Vacational tourism makes possible and promotes   mutual cultural understanding and a culture exchange   between native populations and guests.

The enquiry covered methods of social research:

a)   Intensive interviews with local inhabitants in all test areas, and with 20 "second-home" owners in Breil/Brigels.
b)   Active observations (visits to club events, church   and local political gatherings).
c)   A written questionnaire to the tourist offices.
d)   Analyses of tourism guidelines and touristic marketing concepts.

Among the study's general claims regarding touristic demand was that the cultural aspect of a region is not one of the primary factors for taking a vacation there. Also, attention paid to culture at the resort is not the tourists' main occupation. Regarding communication and mutual cultural exchange, interests and readiness are individually different. This leads to the third working hypothesis that may be stated by the following modified version (Bücheler, 1990a):

"Vacational tourism is a possibility for mutual cultural exchange. However, it does not necessarily have to take place and it is neither planable nor programmable."

The interviews in Breil/Brigels have shown there is an increased effect of mutual cultural exchange on the conscious identity. With increased tourism at a resort the native population tend to act more self-assured and keep more distance from visitors. (Bücheler, 1990a):

"With the material dependence on tourism the immaterial and spiritual influences have a twofold effect. One is that the hospitality declines. The other--parallel with the takeover of the tourists' immaterial values--is a strengthening of their own material and immaterial values and identities."

In particular, the influence of tourism on the language, religious values (frequently in relation to customs) and on the time budget of the native population has been proven. In all test areas the local dialect seems to be endangered, and tourism furthers the spread of the German language. The influence of tourism on social and religious values is partially counteracted by the Catholic Church with its dogmatic strengths. The influence of tourism on seasonal and daily time schedule of the native population is distinct. In private and in public life there are signs of a faster rhythm in life. There is little time for private conversations, handicrafts or club activities.

An overall observation of the socio-cultural transformation of the mountainous region leads to the conclusion that this transformation is due less to direct touristic factors than to indirect touristic and non-touristic factors, although the effect of individual factors cannot be clearly separated as they influence each other. The mass media (especially television) is one of the latter factors, as is the mobility of mountain populations made available by motorization. Therefore the first working hypothesis can be more precisely summarized as follows (Bücheler, 1990a):

"Tourism is currently one of several factors that determine the culture transformation in mountainous regions. Frequently tourism strengthens or accelerates the cultural transformation. At the same time there is a danger of culture levelling through tourism. Tourism can diminish or enrich cultural diversity. Touristic influences on the cultural diversity must in every resort and in every region be judged on their own merits."

According to the study, touristic influences on the cultural diversity and identity of the native populations were based mainly on the following local factors:

1. Speed of touristic development of a resort.
2. Type and property structure of accommodation offered (hotellery, para-hotellery and collective accommodation).
3. Guest structure.

164

4. Relation of the present tourists in peak season (and therefore the accommodation offer) in proportion to the native population.
5. Seasonal frequency distribution.
6. Types of touristic activities.
7. Personal behavior of the guests and the local inhabitants.

Possible strategies for the maintenance and promotion of cultural diversity and identity of the mountain population have to hinge on these factors. In principle the following generalizing recommendations can be made (Bücheler, 1990b):

. The local inhabitants should utilize their cultural peculiarities and their cultural creativity in an original and imaginative way for tourism. Selected target groups are to be consistently addressed with cultural and other touristic offers.
. Individual persons (creative artists, craftsmen, farmers' wives, etc.) must be won over for cultural work in tourism.
. Funding by public authorities for culture must be generally increased.
. New impulses and unconventional creative artists are to be accepted with tolerance and idealistic generosity.
. Village-political development problems should be settled in open discussions.
. The cultural diversity of the mountainous region must be made accessible to the tourists by regional cooperation.

Another research report on a harmonized touristic development (Seiler, 1989) attempts to evaluate the development of tourist resorts by means of several linked indicators in order to deduce firm causes of action. For this purpose seven target sectors were selected and for each sector indicative threshold values of critical danger limits were assigned.

Regarding the maintenance of the cultural diversity and identity of the local residents in tourist regions the danger limits for all target sectors 1, 2, 4, 6 and 7 are more or less directly relevant. The following measures were proposed for the harmonic development of these target sectors (Seiler, 1989):

. Restriction of the building area: preservation of local character, conservation of nature and countryside.
. Strengthen the agriculture: promotion of the partnership cooperation between agriculture and tourism.
. Promotion of hotellery: new orientation of the building sector (maintaining the traditional building substance); establishing a binding final building phase of the tourist resort.
. Control over land: preference given to local residents over outsiders.
. Care for the indigenous culture: improvement of relations between local residents and tourists.

TABLE 8.5   INDICATIVE THRESHOLD VALUES AS DANGER LIMITS OF
TOURISTIC DEVELOPMENT

Target sector	Danger limits
1 Landscape	A possible doubling of the village size within less than 50 years.
2 Agriculture	A more than 2% reduction of land for agricultural use in five years.
3 Accommodations/ transport	An hourly schedule for cableways and ski-lifts of over 400 meters altitude per overnight guest.
4 Accommodation	A ratio of vacational and second-home beds to hotel beds of more than 3:1.
5 Rate of utilization	A rate of utilization of cableways and ski-lifts in winter of less than 30%.
6 Self determination	A property ratio by local inhabitants to vacational and second homes to not less than 50%.
7 Cultural identity	A ratio of beds to local inhabitants of less than 3:1.

Source: Seiler, 1989

8.4   Successful examples of tourism and recreational planning

The targets and measures of tourism and recreational planning can only be undertaken in Switzerland at the lowest political level--the Communities. The consequences of this are that at national or cantonal (state) level there are no exemplary and *realized* tourism and recreational plans to be found. At these levels it is more a question of factual problem solutions on a limited area setting small signposts for the new orientation of tourism and recreational policy in Switzerland.

8.4.1   Recreational planning Türlersee

a) Point of departure
The Türlersee area is located 15 km from the City of Zurich and is accessible on streets and footpaths. The lake and lakeside have been placed under the Canton of Zurich's conservation order since 1944. In summer the lake is available for bathing and in spring and fall the lakeshore and near vicinity make very popular for walkers. A footpath encircles the whole lake. Benches for relaxation are placed at the most beautiful viewpoints. In the course of time a campsite and a swimming pool have been constructed at Türlersee by private enterprise. For the use of both these facilities a fee had to be paid. For camping and car-parking at the east and north shore of the lake (in private hands) a fee was also charged. The few public car-parks are located away from the lakeshore.

b) Problem

With the growth in the population's motorization the lake area became ever more popular by those seeking recreation. They used the entire lakeside as a bathing area and cars were parked as near as possible. Even in 1968 on fine summer Sundays there were 1,400 cars and more than 3500 persons enjoying the Türlersee (Jacsman, 1969). Swimming on fine weekends led to an excessive burden on the lakeside where part of the reed beds (rare marshy biotopes) and agricultural land were negatively influenced. Development of bathing activities increasingly placed the Cantonal conservation order under question.

c) Solution

In 1973, to overcome the conflict, a "Türlersee working group" was founded. It devised proposals for order the leisure activities in the Türlersee conservation area.

The project was realized from 1978 in phases. The most important measures of the concept (Swiss Tourist Board, 1981) were:

. Providing 4 public car parks for 320 cars in addition to the already existing 700 private car parking lots (swimming pool, restaurant; figure 8.9).
. Enforcing a parking restriction on all streets, and requests to the private residential owners not to permit "wild" parking on their grounds.
. Police surveillance on the parking law on peak days during the swimming period. When all car-park facilities were full, the erection of signposts to the approach roads.
. Provision of public leisure areas (apart from the private swimming pool and camp site). The remainder of the lakeshore that was suitable for camping was cordoned off with fencing for the conservation of the reed belt.
. Restoring and partially rebuilding the lakeside circular footpath. Enforcement of a driving and riding prohibition on this path.
. Facilities at leisure areas for recreational activities and games within the vicinities of the car-parks.
. Erection of information notices about the use of the conservation area.
. Organization of the maintenance of the recreational and conservation area (waste disposal, care of the bog meadows, etc.).

Even after a few years it was found that the Türlersee recreational concept had proved successful. Overstrained nature began to regenerate and the interests of recreationists and farmers were fully safeguarded.

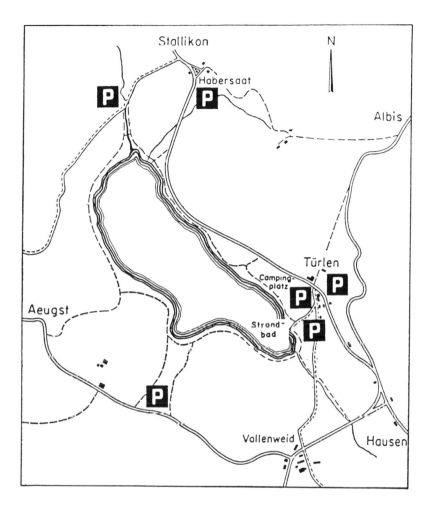

Figure 8.9   LAKE TÜRLER AND ENVIRONMENT WITH NEW PUBLIC
CAR PARKING

Source: Schweizer Fremdenverkehrsverband, 1981

8.4.2   *The concept of nearby leisure--without automobiles*

a) Point of departure
Currently, more than 70% of all day tourists in Switzerland use their own
automobile (Schmidhauser, 1973). The motorized recreationists flow to the very last

corner of natural countryside and impair the very qualities that are sought: back-to-nature feeling, peace and clean air. A working group--"Countryside protection and leisure traffic"--has worked out a concept with whose assistance the authorities, planners, leisure traffic organizers and recreationists, were given the opportunity of an attractive close proximity recreational area without having to resort to their own vehicles. "Within the concept the needs for close recreational facilities are to be satisfied and improved. This can only be attained if the adaptation to the countryside and the interests of the local population are not neglected, so that neither the ecological nor the socio-economical impact limits are exceeded" (Aerni et al., 1980).

b) Problems and Targets

Recreational areas free of automobiles are very attractive for recreation. Therefore, countryside areas have to be created without unnecessary automobile traffic. In addition, the needs of the local inhabitants are to respect. Important tasks emerge in context with the inside and outside development and with the channelling of the through traffic. The recreation areas must be easily and quickly accessible by public and private transport. Car-parks must be provided at the periphery of the recreation areas. These recreation areas must be provided with a properly marked network of paths that can be used by leisure seekers on foot, on horseback or by bicycle, and also by agricultural and forestry traffic and local residents. For the larger valleys and hill ranges an existing well-constructed street for traffic is to be reserved. In a closely populated country such as Switzerland, automobile-free leisure areas are not easy to plan. In order to show the feasibility of the concept the working group "Countryside protection and leisure traffic" attempted to put their ideas in a definite example. The choice fell on the Oberaargau/Emmental/Lakes of Thun/Brienz area in the Canton of Berne.

c) Solution

The planning began with an inventory of the footpath network, facilities for meals and accommodation, and public and private traffic routes. Based on the existing situation the main footpath routes were defined that connected the larger places on the outskirts of the recreation area. These places have an express train station, car-parks, certain accommodations and catering and an information center. The main hiking routes enable walks of one or more days right across the recreation area. In a subsequent working stage the suitable places for stops and secondary places of departure for the main hiking and walking routes in the recreation area were defined. At these places a minimum offer of beds, catering facilities and car-parks were necessary. In a third working step the round-tour footpaths and walking and hiking paths in the region were established. Parallel to the arrangement and classification of the footpaths the classification of automobile-free zones according to the basic layout was carried out. The conflict between footpaths for walkers and

roads for motorized traffic, and the conflict between the various recreation forms
with basic uses (agricultural and forestry, nature conservation and residential areas)
had to be solved. The realization of the vehicle-free recreation area was undertaken
in cooperation with regional and local planning, cantonal and local authorities and
in connection with the development concepts for the mountain region. Although the
vehicle-free recreation area in the Oberaargau/Emmental/Lakes of Thun/Brienz
could not be fully realized, the first results of the aims of the working group may
be seen in a positive light (figure 8.10).

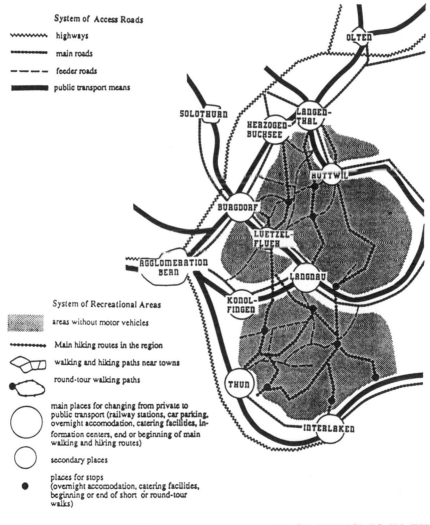

Figure 8.10   RECREATIONAL AREAS FREE OF MOTOR-VEHICLES IN THE
REGIONS OBERAARGAU/EMMENTAL/BERNESE OBERLAND
WITH MAIN WALKING AND HIKING ROUTES

Source: Aerni et al., 1980

*8.4.3   The process of "concentrated building areas" in local planning of the
        Sils/Segl and Silvaplana communities*

a) Point of departure

The Upper Engadine lakeside area and its surrounds constitutes some of the most
beautiful mountain scenery of the Canton of Grisons, with strong glaciers and an
unique high altitude lake valley. The lakes are partly in the area of the communities
of Silvaplana and Sils/Segl. Especially fascinating are the plains between the lakes:
the plateau between the Lej Segl and the Lej da Silvaplauna in the community of
Sils/Segl and the region between the Lej da Silvaplauna and the Lej da Champfer
which, with the bordering valley lands, is owned by the Community of Silvaplana.
The lakeside scenery was already recorded in the 1963 inventory of landscapes and
natural reserves of national importance. In 1972 the government of Canton Grisons
enforced a law on the conservation of the lakeside area. The Upper Engadine region
with its center of St. Moritz is internationally famous as a tourist center. With the
rise of winter tourism in postwar years, the tourist resorts developed rapidly and,
parallel to the expansion of touristic transport facilities and the ski region,
parahotellery, and particularly the building of second homes went through a boom
phase.

b) Problem

Figures 8.11 and 8.12 show the building zones in the Communities of Sils/Segl and
Silvaplana on the basis of the old building regulations of 1962 resp. 1970. The
planned conventional building area took up a great deal of land. The most crucial
planning principle was the boundary distance between buildings. The building along
the lakeshore according to the old building regulations would not only have
diminished the valuable agriculture land in the valley plains, but also strongly
impaired the countryside of the Upper Engadine lakeside.

c) Solution

In the seventies the local planning regulations of the Communities of Sils/Segl and
Silvaplana were revised. The main aims of the new local planning policies were
(Steiger, 1976):
- To preserve the countryside by restricting the building on the reduced areas
  available for building.
- To maintain the Engadine "atmosphere" by keeping to local requirements in new
  buildings.

    To reach these aims the new local planning policy was obliged to adopt the
procedure of "concentrated building areas". This method has nothing to do with
technical design criteria such as shape of the building plots or boundary distancing,
but rather endeavors to keep many open surfaces together. To this end the

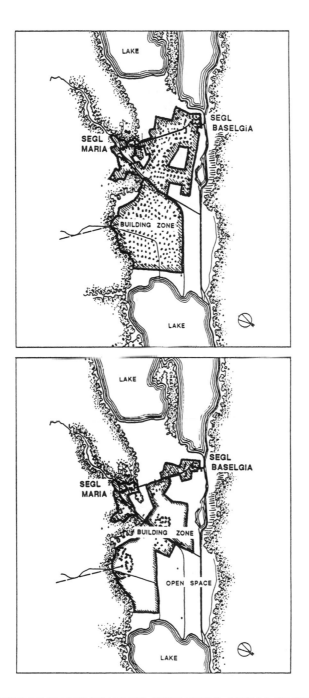

Figure 8.11   THE "CONCENTRATED BUILDING AREA" IN THE FRAME-
WORK OF THE LOCAL PLANNING OF SILS/SEGL

Source: VLP, 1976

172

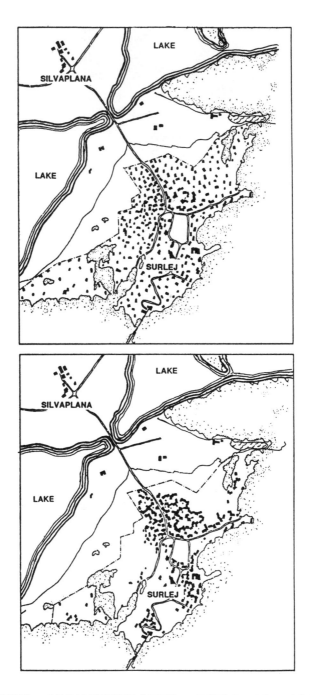

Figure 8.12   THE   "CONCENTRATED   BUILDING   AREA"   FOR
SILVAPLANA

Source: VLP, 1976

previously planned building volume was concentrated on a fraction of the original building zone area. Every landowner was allotted three plots. On two of these he was free to erect the same building volume as he would have been able in the normal type of building. The third plot remains free and can be used for agricultural purposes. The building prohibition was entered in the land registry as servitude in favor of the Community. An additional advantage of the concentrated building area was in the lowering of development costs. The disadvantage of the building restrictions was reduced in that the planning of the total concept is binding for all who want to build and is already known at the time of buying the land. The "concentrated building area" has greatly contributed to a permanent protection of the threatened Upper Engadine lakeside scenery without the need to pay compensation (Weiss, 1981; figures 8.11 and 8.12).

8.4.4   The Alpine rest zones in Pontresina's land planning

a) Point of departure
The Pontresina Community is also located in the Upper Engadine. And it also has an international reputation as a tourist resort. Moreover, the leisure facilities in winter are one of the most important touristic offers of the Community. Consequently in Pontresina too more and more areas of the mountainous region were being exploited for winter sports.

b) Problem
The cableways, ski-lifts and the prepared ski-runs burdened the countryside and natural habitat of the flora and fauna, worsened the danger of erosion and reduced the crops of the mountain agriculture. It also transformed a previously almost intact, natural countryside into a civilization scenery estranged by technology, which could no longer guarantee the natural and countryside ambience which tourists seek in summer. For this reason, the demand to halt the technological march on the mountainous region or at least to considerably limit it stems not only from the protectors of nature and countryside. Local planning offers possibility to fulfil these demands.

c) Solution
The main task of local planning is to regulate the use of the soil by a binding settlement of the zones for all landowners. The "zones utilization" regulates the permitted use of the soil. The Pontresina Community has created in its local planning "special utilization zones" for the development of cableways and ski-run construction (Building regulation Pontresina 1988). The "Winter Sports Zone" covers that land necessary for the exercising of winter sport such as ascent and downhill skiing areas, routes for transport facilities, surfaces for mechanized snow-laying of ski-runs, exercise areas, cross-country ski-tracks and sledge runs.

The "Alpine rest zone" is dormant during the winter sports season. It covers a wide region of special areas especially worth conserving. Mechanized touristic transport facilities and motorized traffic is not allowed in the rest zone. Excluded in this regulation is the use by agriculture and forestry. The Pontresina Community introduced the zone plan (with connected protection zones in which touristic development facilities are forbidden) when the local planning policy of 1988 was being revised.

8.4.5   Travel offered by the PTT for nearby recreation

a) Point of departure

As already mentioned, the nearby recreation in Switzerland hinges to a great extent on the private motorcar. One glance at the railway and bus timetables is enough to show that most leisure areas close to built-up locations are within easy reach by public transport. The reason for preferring private automobiles is that public transport is not considered to be of sufficient quality. Recently, therefore, exponents of public transport have been improving the appeal of their range through targeted campaigns and special offers. In Switzerland the Swiss PTT (Post, Telephone, Telegraph Organization) offers a wide range with their public post-auto coaches (coach lines).

b) Problem

A grave handicap for using public transport was the lack of information available for those seeking recreation. Leisure seekers generally did not know how often and where the journeys went to and how they could combine the journey with their recreational activities. Moreover, the price of the ticket - especially for large families - could be another reason for not using public transport. Finally, the conveyance of equipment for leisure and sports is often very difficult in public transport.

c) Solution

The PTT offers to the area, with consideration of the legal, organizational and economical conditions, special packages for summer and winter seasons. The range is summarized in a brochure, which passengers in all built-up areas receive in their home mail. The brochures are also available from all post offices and at all information centres of the PTT and public transport offices. The range is generally as follows:

. Block arrangements in winter (Postbus and day ticket for transport facilities in the ski area).
. Combined excursion tickets (Postbus, train, cableways, boat).
. Postbus walking ticket (especially for walks, see figure 8.13).

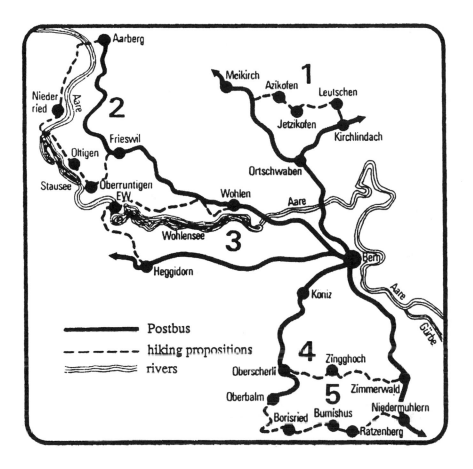

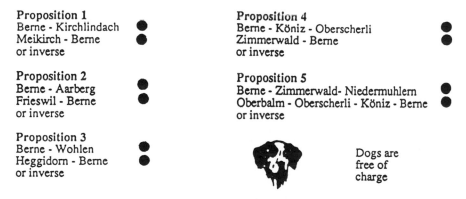

Hiking route propositions

Proposition 1
Berne - Kirchlindach ●
Meikirch - Berne ●
or inverse

Proposition 2
Berne - Aarberg ●
Frieswil - Berne ●
or inverse

Proposition 3
Berne - Wohlen ●
Heggidorn - Berne ●
or inverse

Proposition 4
Berne - Köniz - Oberscherli ●
Zimmerwald - Berne ●
or inverse

Proposition 5
Berne - Zimmerwald- Niedermuhlern ●
Oberbalm - Oberscherli - Köniz - Berne ●
or inverse

Dogs are
free of
charge

Figure 8.13   OFFER OF PTT (POSTBUS SERVICES) FOR HIKING TOURS
IN THE REGION OF BERNE

Source: Schweizer Fremdenverkehrsverband, 1981

The journeys are usually carried out on scheduled routes. When demand is heavy additional coaches can be arranged. The current response is thoroughly satisfactory. The demand for block arrangements and special excursion tickets in winter are showing an increased tendency. Disadvantages are the long travelling times of the scheduled routes and the still unattractive price of the tickets.

REFERENCES

Aerni, H. et al., 1976. Eine motorlose Freizeit - Schweiz. Ein Leitfaden für die Ausstattung und Gestaltung von Erholungsgebieten mit besonderer Berücksichtigung motorloser Aktivitäten. Herausgegeben von der Schweiz. Stiftung für Landschaftsschutz und Landschaftspflege, Bern.

Baugesetz Pontresina (Revision 1988).

Beratende Kommission für Fremdenverkehr des Bundesrates, 1979: Das Schweizerische Tourismus-Konzept. Grundlagen für die Tourismuspolitik. Schlussbericht. Bern.

Bücheler, R. and H. Elsasser, 1987. Tourismus und kulturelle Vielfalt. Zu einem Projekt im Rahmen des Nationalen Forschungsprogrammes "Kulturelle Vielfalt und nationale Identität" (NFP 21). In: Dokumente und Informationen zur Schweizerischen Orts-, Regional- und Landesplanung DISP, Nr. 88.

Bücheler, R., 1990a. Soziokulturelle Aspekte im Tourismus: In: Die Region. Nr. 1/90.

Bücheler, R., 1990b. Kurzfassung. Kulturelle Vielfalt, Tourismus und Identität im Berggebiet. Sammelband NFP 21. Basel.

Bundesgesetz über die Raumplanung vom 22. Juni 1979.

Elsasser, H. and H. Leibundgut, 1984. Kleine Fremdenverkehrsgeographie der Schweiz. Wirtschaftspolitische Mitteilungen, 40. Jg., 3 Heft. Zürich.

Elsasser, H. et al., 1982. Nicht-touristische Entwicklungsmöglichkeiten im Berggebiet. Schriftenreihe zur Orts-, Regional- und Landesplanung, Nr. 29. Zürich.

Jacsman, J., 1969. Erholung am Türlersee. Arbeitsberichte zur Orts-, Regional- und Landesplanung, Nr. 8. Zürich.

Jacsman, J., 1990. Die mutmassliche Belastung der Wälder durch die Erholungsuchenden. Eine makroanalytische Studie zur Schätzung der Nutzungsintensitäten der Walderholung in der Schweiz. Berichte zur Orts-, Regional- und Landesplanung, Nr., 79. Zürich.

Kaspar, C., 1975. Die Fremdenverkehrslehre im Grundriss. St. Galler Beiträge zum Fremdenverkehr und zur Verkehrswirtschaft. Reihe Fremdenverkehr, Band 1. Bern und Stuttgart.

Krippendorf, J. et al., 1986. Freizeit und Tourismus. Eine Einführung in Theorie und Politik. Berner Studien zum Fremdenverkehr, Heft 22. Bern.

MAB 1976: MAB-Information Nr. 3. Eidg. Amt für Umweltschutz (Hrsg.). Bern.

Messerli, B. and P. Messerli, 1978. MAB Schweiz. In Geographica Helvetica, Nr. 4.

Messerli, P., 1988. Mensch und Natur im alpinen Lebensraum. Risiken, Chancen, Perspektiven. Bern, Stuttgart.

Monteforte, M. et al., 1973. Tourismus. Entwicklungstendenzen im Tourismus und operationale Planungskonzepte. Teufen.

Müller, H.R., 1986. Tourismus in Berggemeinden: Nutzen und Schaden. Eine Synthese der MAB-Forschungsarbeiten aus tourismuspolitischer Sicht. Schlussberichte zum Schweizerischen MAB-Programm, Nr. 19. Bern.

Müller, H.R., 1987. Tourismus in Berggemeinden - ein vernetztes System. Touristische Vernetzungsmatrix: Ein methodischer Ansatz für eine umfassende Schaden-Nutzen-Analyse. In: Dokumente und Informationen zur Schweizerischen Orts-, Regional- und Landesplanung DISP, Nr. 87. Zürich.

Schmidhauser, H.P., 1973. Der Wochenendausflugsverkehr in der Schweiz. Institut für Fremdenverkehr und Verkehrswirtschaft an der Hochschule St. Gallen. St. Gallen.

Schweizerischer Bund für Naturschutz u.a., 1967. Inventar der zu erhaltenden Landschaften und Naturdenkmäler von nationaler Bedeutung. Zweite, revidierte Ausgabe. Basel.

Schweizerischer Fremdenverkehrsverband, Dokumentations- und Beratungsstelle (1981): Der rollende und ruhende Verkehr in Naherholungsgebieten. Bern.

Schweizerischer Fremdenverkehrsverband u.a., 1981-1989. Schweizer Tourismus in Zahlen. Bern.

Seiler, B., 1989. Kennziffern einer harmonisierten touristischen Entwicklung. Sanfter Tourismus in Zahlen. Berner Studien zu Freizeit und Tourismus 24. Bern.

VLP Schweizerische Vereinigung für Landesplanung, 1976. Ortsplanung und Landschaftsschutz in Kurorten. Schriftenfolge 19. Bern.

VLP Schweizerische Vereinigung für Landesplanung, 1976. Wirkung und Nutzen der Ortsplanung in Ferienorten. Schriftenfolge 27. Bern.

Weiss, H., 1981. Die friedliche Zerstörung der Landschaft und Ansätze zu ihrer Rettung in der Schweiz. Zürich.

CHAPTER 9

Managing the resort community planning and design process

by
Joe Porter
Principal in Design Workshop
Denver, Colorado, USA

9.1   Introduction

The purpose of this paper is to share some observations and experiences about the design of resort communities and the community development process for the purpose of improving the quality of design of resort communities. The paper concentrates on design, to complement the attention given to planning and policy of the other papers in this text, with the additional objective of furthering good community design as an important issue for educators, elected officials, and practitioners concerned with human use of the physical environment.

The term community design is no more than Frederick Law Olmstead's definition of landscape architecture which is "The design of the land and the objects upon it". Specifically, community design refers to any element or groups of elements which make up a community, and which are designed with regard to community and environmental context. Community design can include landscape architecture, site planning, site design, urban design, engineering and architecture, provided they are practiced together and in a community context.

Although the process put forward applies to the design of any resort or non resort community, mountain resort communities are the ideal medium to examine community development concepts. They are of a size and scale that can easily be monitored, and their typically isolated locations reduce variables. Mountain resort communities are likely to stress environmental or social values which require design exploration and other-than-normal design solutions. In addition, the economics associated with successful resort development are more likely to support a higher level of financial and management resources as applied to the design process.

There are three basic themes in this paper. The first provides a brief summary of the changes in the design professions and particularly landscape architecture in the United States during the past 25 years, and the importance of re-establishing design as a priority.

Second, the community development process is discussed to create an awareness of the scope of community development and the leadership role that design should play in the process.

Third, this paper discusses the importance of design and design management in resort community development.

This paper is grounded in experiences gained through the author's consulting work in designing both permanent and resort communities. The suggestions relating to design, design management and application of computer technology are techniques which Design Workshop, the author's firm, currently incorporates in private practice.

9.2   State of design

In practicing landscape architecture in the context of community development, design as it is, based on traditional community values, is playing a less significant role in determining the form of communities. This is somewhat less true for resort communities, which constitute some of the best recent examples of new community development in the United States.

This situation is created because the problems associated with community design are becoming more complex and unmanageable, and the design professions' abilities to participate in a comprehensive leadership position have not kept pace. Although there are exceptions, the trend seems to be for the professions of architecture and engineering to become more focused on design of individual community elements and the profession of landscape architecture to become less focused on design.

The reduced focus on design in landscape architecture is understandable. The profession has always been diverse and quick to embrace new community and environmental issues. The profession (and design as a priority) changed significantly with the environmental and social revolution of the 1960s and the 1970s. The changes were necessary and positive and in fact the need for change still exists. One of the costs, to a relatively small profession, has been a dilution of energy directed toward design.

While teaching from 1965 to 1974, I observed faculty interests turning from design, to broader environmental and social concerns. Students, particularly at the graduate level, began taking a secondary interest in the traditional forms of practice to become planners, managers and environmental scientists. The scope of landscape architectural education has been expanded since that time and with funds growing more limited individual programs today are forced to address a wide variety of subjects with limited resources. The result has been a trend toward a more general professional education and away from a priority of problem solving and design.

If our firms' experience is any indication, professional practice has gone through a similar evolution. Our firm was founded in 1970 with the objective of bringing design professionals together to design without professional barriers. Our work product was highly visual, took place on large throw tables, consumed enormous wall space, and models and sketches were everywhere.

Responding to issues of the time we attempted to be responsive to environmental and social concerns. Many of our business opportunities were development projects with special approval problems due to the public's environmental concerns. This was an appealing market because of the enormous contrasts between the problems and potential of the development industry. At one extreme, developments were destroying irreplaceable sensitive environments. In contrast, such private entities as the Rouse company in Columbia, Maryland, were using community development as a tool to conserve the environment, to re-establish neighborhoods, to innovate in the health care delivery, and to lead racial integration in the Baltimore suburbs.

For ten years we concentrated on issues other than design. Our professional growth was directed to broadening our knowledge through collaboration with professionals in all areas of community development. We worked hard to under-stand the development process, all the while become more process-oriented and playing a larger role in project coordination. Our work products evolved from large drawings and models to reports in a written format.

During this necessary and valuable period, our design skills atrophied, requiring a dedicated effort to retrieve and re-develop design and implementation skills capable of achieving quality in community design.

I believe our experience is indicative of a trend in much of the profession. The time has been well spent, but now it is critical for some in the profession of landscape architecture to return to an emphasis on design as a principal tool in community development.

The need for quality community design is more important today than it was 25 years ago. A good case can be made that in the early 1960s the abilities of the design professions exceeded society's expectations. It was a simpler time when projects would later be criticized for lack of substance and responsiveness to issues we understand today.

182

The situation is now reversed and the substance of community development problems and society's expectations exceed professional design abilities. It's extremely important that we translate the values and information of the past twenty-five years into built communities.

9.3   Community development process

The physical forms of communities are not being created through design but are evolving through incremental decisions borne of a fragmented development process. This process determines community form with a force that is beyond the control of the designer. Even mountain resort communities, which are smaller and more manageable and which provide some of the best examples of comprehensive community design, until recently, have not been designed. Formation of a strategy to design communities comprehensively, we now know, requires an understanding of the scope of the community development process and where design decisions take place in the process.

The Community Development Matrix (see figure 9.1) provides a way of simplifying and looking conceptually at community development to understand the characteristics of the process and to clarify the relationship between design and other community development decisions.

Community development isn't as complicated as it is confusing. Several obvious factors account for this confusion. As the matrix illustrates, the scope of community development is all encompassing and includes decisions about every element of community life and environment. Although the matrix shows a well ordered decision sequence, in reality, decisions occur out of order and simultaneously, and are in a never ending state-of-change. There is no beginning or end to the process. The variables are infinite and a decision at any point illustrated on the matrix can ripple through the entire process.

The purpose of the matrix then is to freeze a concept of community development, to begin to understand the process and identify where community design can play a more significant role, exert greater influence, and accomplish more for the effort.

The decision levels in the matrix describe the hierarchy of development decisions which are made in community development. These decisions, which range from policies on the national scale to the detail design of street furniture, are all part of the community development.

Development topics	Decision level	National	Regional	Area	Village	Building cluster	Building
Environment							
Program							
Transportation							
Utilities							
Economics							
Legal							
Social services							
Politics							
Planning							
Design							
Marketing							
Operations							

Figure 9.1   COMMUNITY DEVELOPMENT MATRIX

The decision line represents decisions made in the course of community develop-ment. The subjects listed represent only a few of the primary categories because the actual list is endless. Although the list of decisions is far from complete, the matrix begins to illustrate the comprehensiveness of community development.

Each square symbolizes a community development decision made at a time and point in the process and based on a finite body of knowledge. Community develop-ment at its best would have the activities between squares fully coordinated and integrated. The opposite however is true as the nature of the process creates a natural isolation between the professionals and information represented by the squares.

This isolation occurs partly because in this age of specialization, the squares represent decisions made by technologists with expertise in a single area. Much of the progress made in community development in the past 25 years has been in individual areas of technology, such as the environmental sciences, economic modeling, finance, construction management, market analysis, advocacy participation, and all of the areas of impact analysis. It seems that as the body of knowledge grows, interaction, not to mention integration, becomes more difficult.

The process is equally polarized by special interests. The free enterprise system in the United States creates an opportunity for individuals to profit from a wide array of development, building, business and consultant activities, not all of which are in the interest of good community.

In addition, national, state, and local laws and politics provide the public a significant say and legal recourse in governmental approvals of mountain resort development activity. These rights also are not always used in the best interest of good community.

As a result of technical focus and special interests, the lines on the matrix represent major communication barriers which are indigenous to the community development process. Community development issues are further isolated by the management technique used to manage the cumbersome process, which is to isolate and control the individual decisions on the matrix. An engineer who was a project manager for one of our major clients once compared the process of managing the relationship between our firm and an architect to training a puppy by hitting him on the nose with a newspaper. So it goes with managing design professionals.

The matrix was originally diagramed to understand the scope of community development and our firm's capabilities relative to the process. However, over the years the diagram has been the basis for establishing professional development priorities for the firm. It has been a teaching tool used to explain community design to staff, clients and members of development and planning teams.

The diagram also has been used in managing projects. An example is Baqueira/ Beret, a ski resort in the Valle de Aran (Valley) in the Spanish Pyreneese. The project was to significantly expand the existing resort of Baqueira with the addition of a new village, Beret. The program was to set up a planning and design process, design the new village with significant consideration of the existing Village of Baqueira and the Valley, and to teach the planning process to the company's staff so they could continue the process of planning and implementation.

The matrix (see figure 9.2) was used to introduce the planning process to the developer and to establish the project's initial scope and procedure . It then became the structure and format for the work program, and was used as the agenda for intense work sessions in Spain (see also figure 9.3).

In the course of preparing the Beret Plan, numerous planning issues and implementation strategies were addressed for the existing village, other resort sites, the Valley, and adjacent watersheds. The matrix provided a system to take issues as they arose, systematically organize them, and take action--no small feat given the comprehensiveness of the development, the introduction of a new process, and the language barrier.

9.4   Design management

Design of mountain resorts was much simpler twenty-five years ago when such complexes as Vail and Snowmass were coming into existence. There generally was less sensitivity to the environment and other contextual issues, and the development industry was less sophisticated and experienced. Resorts were typically one-dimensional, focusing around a single activity, usually skiing, with little consideration of community issues.

People were attracted to existing communities like Aspen because they loved the place. Once there they began to build amenities and community facilities. "We started a ski company and built a lift on Aspen Mountain because we got tired of walking uphill to ski", says Darcy Brown one of the founders and long time president of the original Aspen Skiing Corporation.

Some areas like Aspen were fortunate to have benefactors like the late Walter Papke and R.O. Anderson, who established cultural institutions like the Aspen Institute and the Aspen Music School.

The few resorts designed professionally were initially conceived by engineers or architects. Such individuals were elitely known as "The Designer", and were relatively free to design what they thought appropriate. Most new resorts were lacking with the notable exception of the original Vail Village, in Colorado. European resorts of the period tended to be contemporary architectural statements totally inappropriate to the mountain environment.

The public concern for the environment which surfaced during the 1960s and 1970s, and the intense competition for tourism, served as a wake-up call for the mountain resort industry which has grown in size and sophistication over the past quarter century. It has become clear that people care about their community

Development topics	Decision level	Spain	Valle de Aran	Baqueira/Beret	Beret	Neighborhood	Building
Environment							
Program							
Transportation							
Utilities							
Economics							
Legal							
Social services							
Politics							
Planning							
Design							
Marketing							
Operations							

Figure 9.2   BAQUEIRA/BERET DEVELOPMENT MATRIX

environments with a growing resolve to participate in community decisions and to not let change go unchallenged. This applies to the United States, Canada, Spain and in every community in which we have worked.

Aspen has taken extreme and restrictive positions in protecting existing community values, which demanded even higher design responses to every imaginable community development trend. The development issues which have been confronted in Aspen are being raised by residents and consumers universally. The result is an obvious trend toward quality resort environments, indigenous to their surroundings.

Now, in the 1990s, there are higher expectations for the quality of life. The design problem is much more defined, and excellent information is available about all

V. Baqueira 1700 building cluster prototypes	Transportation - EYSER	Buildability/Engineering - SERELAND	Utilities - SERELAND	Architectural Form	Skiing - Miguel & Larry	Management - Miguel & Larry	Life Quality - Miguel & Larry	Real Estate Marketing & Development - Carlos	Political Feasibility - Jose	Legal - Jose	Economic - Cami	Environmental - Environmental Group	Meteorology - Fernando
1. Program for Option 1	●		●		●	●		●					
2. Master plan concept for Option 1	●				●	●		●					
3. Summarize program issues				●			●	●					
a. building height	●		●			●	●	●					
b. service	●				●	●	●	●					
c. automobile circulation & parking				●			●	●					
d. architectural character					●	●	●	●				●	
e. open space system							●	●					
f. open space character				●		●		●					
g. sun & wind related criteria													

Source: Baqueira/Beret Planning and Design Work Program that demonstrates the Community Development Matrix integrated into project management.

Figure 9.3   EVALUATION MATRIX FOR THE BAQUEIRA/BERET DEVELOPMENT MATRIX elements of the environment, society and the development process. The question remains what can be done in order to design communities which will fulfill our expectations.

Design status

The most successful resorts have resulted when design has enjoyed a status on the development team equal to that of finance, operations and other development activities. This is not a subtle distinction and it occurs when ownership or management establishes and mandates, from the beginning, that design of the environment is the foundation of the business and operation plan.

Under these conditions design is considered a strategic as well as physical decision making activity. Design in this situation is considered a comprehensive process and a long term interactive relationship is established between designer and the resort.

Design management

Successful resort design also results from managing work programs, people and information for the purpose of achieving a quality community. Design management at its best avoids incrementalism and links the components of community development symbolized by the squares on the community development matrix. Design management typically has occurred as a byproduct of managing government approval processes, or as a result of a development manager or owner possessing a strong personal bias toward design. Occasionally, a consultant provides appropriate design management service as part of a design package.

Design management as an independent activity is a relatively new idea which will become a critical discipline in the community development industry. This opportunity exists because design is a part of the community development process which few in development or government are capable of managing.

While many know the language, the cost and the time required to talk to an accountant or lawyer, few can communicate with their designer. Design is such a mystery to communities that it's the only element of development which is not regulated. Even Aspen, which has the most stringent development controls in the resort industry, does not directly regulate the appearance aspect of design, with the exception of design review in the town's historic district.

Design management is critical to maintaining an on-going design process. The following list of planning and design activities (table 9.1) summarizes the individual activities which are repeated again and again in community development. Design management assumes that each of these tasks receives the focus necessary for implementation without compromising the continuity required for each task to build on one another.

TABLE 9.1 PLANNING AND DESIGN TASKS

Site evaluation
Program development, including market analysis
Planning and design
Financial programming
Operational programming
Implementation documents

Design places

Successful designs are those that have been designed for the purpose of creating great places. It seems like this would be the obvious objective in preparing plans for any resort. This is not the case since many resorts back into a development plan through a series of incremental steps and commitments prepared for investor packages, financing packages, zoning, utilities and sales.

Design

In the chaos of all of the information, expectations, laws, economic cycles, management processes and general distractions, there must be someone with the ability to focus on design at key points in the decision process. Design is most valuable when started early in the process as a component of initial strategic planning. Public and private developers who are comfortable with and can manage design don't hesitate to integrate design exploration and testing early in the process when the situation warrants.

There are some misconceptions about the cost of design, which delay design activities until "we get our ducks in a row". The costs of making fundamental design decisions early and at timely points in the development process are relatively inexpensive. The primary costs associated with design come from the notion that the design is a final document that must be right the first time. As a result, the information must be complete and presentations final. The primary costs of design come from the gathering of information, such as surveys form environmental mapping and programs; from presentations including zoning submissions, marketing and public review; and from technical documentation required for project implementation. We estimate that in preparing a village master plan for a resort, less than 15% of the time is actually spent making focused design decisions.

Design early in the process, based on experienced site observations and presented in working format, produces some of the most valuable and timely information which a developer can buy. Few developers, public or private, want to build projects that do not look good or that are inappropriate to a community. Developers have egos and take pride in what they create, and are often trapped by decisions that are made early in the process with inadequate on inaccurate design input. Design studies, visualized throughout the process, allow decision makers to see the implications of decisions and thus avoid costly mistakes throughout a project's development.

Visualization

Design visual tools have not kept up with the magnitude of the communication problem in community development. In the past 25 years, great strides have been made in tools for economic modeling, engineering, and environmental analyses. During the same period, however, there have been no new tools seriously applied to visually communicating design. This in ability to recognize, address and communicate issues of visual importance has added to the confrontation associated with community development and change. The physical form or appearance of change, is in many instances the underlying concern in development controversies.

Few people involved in the development of a project are able to read plans. Neither do they have the experience to visualize a proposed design or change in the environment. Consequently, they go through the process with an understanding of factors, such as costs, impacts, zoning, contracts, schedule, size, volume and dimensions, without understanding, in adequate detail, the project appearance. Opponents to projects, likewise, rely on measurable, legally defensible indicators, such as traffic counts, air quality, and water quality to fight projects which they assume will be ugly.

Thus, quantitative measures of developments, like numbers of units or floor area ratios, are of no value in assessing the appearance, appropriateness or scale of development within a community. And, when the visual products of design are used primarily for marketing tools rather than for analytical purposes, it is testimony to the fact that design professionals have failed to demonstrate that design is a process.

As with most processes in society, one must have faith that going through it will yield an acceptable, understandable, and workable product. That's what design is about, and that's why we must return to it if we are to continue to develop in our most sensitive environments.

CHAPTER 10

Use of nominal group technique in tourism information needs assessment and goals development*

by
Carson E. Watt, Interim Head
Department of Recreation, Park and Tourism Sciences,
Texas A & M University, College Station, Texas, U.S.A.

Turgut Var
Department of Recreation, Park and Tourism Sciences,
Texas A & M University, College Station, Texas, U.S.A.

James C. Stribling
Department of Recreation, Park and Tourism Sciences,
Texas Agricultural Extension Service, College Station, Texas, U.S.A.

10.1   Introduction

The application of nominal group technique (NGT) by the Texas Tourism and Recreation Information Program (TTRIP) began in 1986 when the concept was being developed of TTRIP as a coordinated research and information service for the Texas tourism industry. Early in that process organizers recognized an over-whelming tendency to focus on the technology of information processing and the availability of data bases rather than the appropriateness of those data to benefit the program's clientele.

The improvement in electronic data storage and information retrieval systems has enhanced the potential for more informed, reliable tourism decisions. Additionally, the computer has stimulated the appetite for information at all levels of the tourism and recreation industry. Despite these improvements, community, business and organizational tourism decision-makers continue to labor to acquire appropriate and reliable data with which to make both tactical and strategic decisions. The fragmented, diverse nature of the tourism and recreation industry further complicates the process of determining the specific needs for which information is required. Despite these conditions more can be done to enhance the quality and appropriateness of information. The first step in this process is to more accurately identify the decision-making needs for which information is required.

---

*   This chapter was prepared originally for the International Association of Travel Research and Marketing Professionals, June 19-22, 1988, Montreal, Canada.

10.2   Conceptual framework

It was the conditions mentioned in the previous section that inspired the Texas Tourism and Recreation Information Program to search for more accurate techniques for assessing the information needs of its clientele. Figure 10.1 illustrates the TTRIP conceptual information model. As the model denotes, information needs must be identified through some process before evaluating data for its appropriateness. Several methods of needs assessment were identified. The desire was to utilize a process which would determine <u>what</u> information is pertinent, <u>when</u> it is required and <u>what form</u> is most appropriate to make data useful in decision-making. Also important was determining the "degree of resolution or detail" needed for clientele decisions. It was determined that a variety of methods would be required for TTRIP's needs. Advisory groups, Delphi, nominal group, and surveys were specific methods identified. The challenge was to narrow the focus of clientele to identify their specific concerns for which they needed information. The heterogeneous nature of the tourism industry further requires segmenting information needs. Once these needs are established, specific goals and alternative strategies which will require information can be articulated.

The quality of information is only as good as its relevance with decision-making needs and the technique or combination of techniques used to solicit it. Time, cost, technical know-how, and desired outcomes are criteria that dictate the appropriate choice. For TTRIP, the nominal group technique (NGT) was viewed as potentially appropriate for the program. The underlying reasons for this choice are: a) the paradox of the diverse nature of motives which commonly comprise tourism businesses and organizations, b) the need for collective marketing or product development initiatives for a destination or a region, and c) NGT as a method of group consensus, provides a means by which agreement on achievable goals and other needs can be achieved. Propensity toward group consensus NGT was felt to provide a means by which agreement on needs could be achieved.

10.3   Nominal group technique

Nominal group process is a structured problem-solving or idea-generating strategy in which ideas are generated and combined in a face-to-face non-threatening group situation (Butler and Howell, 1980). The nominal group technique was developed by Delbecq and Van de Ven (1975) in the late 1960s. The process was developed from studies which examined the character of communications between two different groups: (1) a "nominal group," that is one in which members identified ideas through written communication, followed by (2) the interacting group, where members interacted through discussion of these ideas (Gill, 1982). The results of this research showed that the "nominal" group process

195

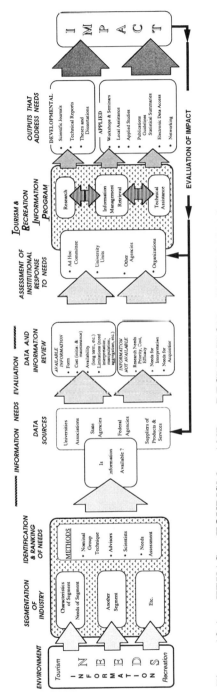

Figure 10.1   TEXAS   TOURISM   AND   RECREATION   INFORMATION
SYSTEM MODEL

generated a higher number of ideas on a given problem, a greater quality of ideas (as determined by external judges), and a set of ideas which were more inclusive than those which were generated by spontaneously interacting groups. The interacting groups were better at the evaluation tasks of weighing alternatives and selecting alternatives which they were able to clarify through interaction. An additional advantage of nominal group is that it can be varied and combined with other group interactive methods (Bartunek and Murningham, 1984; Hegedus and Rasmussen, 1986; Lundberg and Glassman, 1983). Hegedus and Rasmussen (1986) found that NGT, "may make members more open to others' ideas."

A significant benefit of nominal group is that it enables individuals of different ranks in a firm and/or positions in a community to participate since the fear of social pressure or interpersonal conflict is largely removed through the structured process. A sense of involvement and personal contribution is further augmented by the NGT process because it overcomes differences in status, leadership abilities and communication skills.

Steps in nominal group process

There are four basic steps in the nominal group process: (1) silent generation of ideas in writing, (2) the discovery step, or "round-robin" elicitation and public recording of each individual's ideas, (3) discussion for clarification, and (4) ordering of ideas based on common criteria by secret ballot. Table 10.1 provides more detail to each of these steps.

Variations in the nominal group process

As mentioned before, nominal group can be varied to accommodate characteristics peculiar to the community, organization or firm. TTRIP varied the process to accommodate the unique needs of the targeted tourism clientele. The process was not varied however to evaluate the efficacy of one NGT variation over another. Rather it was judged to be most practical for the desired outcomes and time constraints to limit the discussion step and add the steps known as the "elimination of redundancies" and "combining of similarities". This modification is illustrated in Figure 10.2. This report of applications of NGT is meant to provide feedback of an applied nature with reflections on the implications of the results.

TABLE 10.1   STEPS IN NOMINAL GROUP PROCESS

STEP 1 SILENT GENERATION OF IDEAS IN WRITING

The question or issue is written on a flip chart for all to see. The phrasing of the question is perhaps the most important point in the whole NGT process. It must be pretested and carefully written or else the subsequent effort is wasted. The group members are requested to write down their ideas without engaging in any discussion with other group members. Seven or eight minutes are allotted for this activity. The ideas should be written in short, brief phrases, not long sentences, because they all have to be recorded by the group leader in Step 2.

STEP 2 ROUND-ROBIN RECORDING

Each group member is asked to read out one of their ideas, in turn. As each idea is read, it is numbered and recorded on a flip chart visible to all group members. No discussion of the ideas is permitted at this stage, but participants are encouraged to "hitchhike," that is, to continue writing down ideas, if those of their colleagues recorded on the flip chart spark new thoughts. This step is likely to take 40 to 50 minutes to complete.

STEP 3 DISCUSSION FOR CLARIFICATION

This is the first opportunity which individuals in the group have to discuss their ideas. After an hour or so of reflective thought and reading, most groups are ready for discussion. Each item is addressed in turn going down the list. The purpose of this step is to clarify the terse phrases that have been previously recorded. Group members are encouraged to ask questions or make comments about each idea's importance, clarity, meaning or underlying logic. Illustrations expanding on the idea from all members of the group are often helpful. Individuals who suggested an idea do not need to comment on it if they do not want to do so. Heated debate of an idea's merit should be avoided. Typically, this third step of discussing each item in order takes about 60 minutes to complete.

STEP 4 PRELIMINARY VOTE

In this stage, a subset of priority items are selected from the total common list by each individual. Each group member is given seven 3 x 5 cards and requested to write down the best seven ideas on each card. (The group leader should determine the number of ideas to be chosen. A number between five and nine usually works well). The idea number from the master list should be recorded in the upper left-hand corner of each card.

Next, after each participant has silently recorded the seven ideas, they are requested to rank each one by assigning a 7 (assuming only seven ideas were selected) to the most important one, a 6 to the next most important idea, and so forth, until all have been ranked. The ranks for each idea should be recorded on the lower right-hand corner of each card. At the conclusion of the ranking, the cards are passed to the next person and then to another person who is responsible for reading them out. The intent is to provide anonymity to each individual's ranking.

STEPS 5 & 6 (Repeats Step 3 & 4 -- Optional to reach agreement, if required).

Source: Delbecq et al., 1975.

198

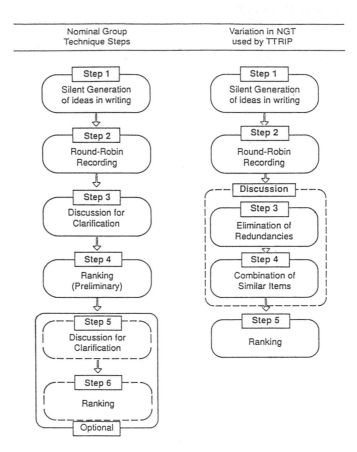

Figure 10.2   FLOW DIAGRAM COMPARING NOMINAL GROUP TECHNIQUE (NGT) AND VARIATION USED BY TEXAS TOURISM AND RECREATION INFORMATION PROGRAM (TTRIP) 1987-88.

10.4   Applications of NGT in tourism information needs assessment and goals development

The nominal group process was applied five times to five different occasions with Texas groups during the period October, 1986 to January, 1988. Four of these cases are described. Most of the preparation and procedural requirements of the NGT process common to all cases are covered in the description of the first case. The description of the remaining three cases covers only variations thought to contribute to the outcome or otherwise enhance the process.

10.4.1 Application #1

Goal. The goal of the first application of NGT was: "To produce a list of prioritized information needs of the Texas tourism industry." It is important to note that the process was externally imposed by the Texas Tourism and Recreation Information Program and the Texas A&M University Sea Grant Program as a means of identifying the information needs of target clientele for educational programs. The project was enthusiastically embraced by the industry participants.

Nature of Group. Representation from a cross section of the Texas tourism industry was sought. Over 60 names were identified by referral, utilizing community, state and industry knowledgeables. The goal was to generate 30-35 participants. Thirty-two representatives of 19 segments of the tourism industry were present for the workshop.

Preparation. As noted previously, advanced preparation is critical to the NGT process. The first task was to formulate appropriate questions to generate focused outcomes. In each of the applications, this step required the most deliberate thought and discussion. Two questions were formulated:

A. What kinds of information do you need, on a routine or recurring basis, in order to make more informed tourism-related business decisions?

B. What kinds of information do you need, in a more strategic sense, for your tourism-related business to get from where you are today to where you want to be in the next 3-5 years?

Each participant received the questions and an agenda prior to the workshop. The correspondence also reiterated the goal and a brief description of the process. A job description for each support position was developed. Support positions included: facilitators (responsible for fostering and controlling group sessions) and tabulators/recorders (responsible for tabulating votes and assisting facilitators during group sessions). A training session was conducted the evening before the workshop for all support positions.

Description of the Process. Five and one-half hours (1/2 day) was scheduled for the workshop. The program began with lunch allowing the morning for travel. As they arrived, participants were given coded badges. The badges were coded to identify three groups and to insure similarity between groups and industry diversity within each group; that is, public and private sector participants were represented in each group. The afternoon agenda was:

SESSION 1. Orientation. (The goal of the workshop was restated and the NGT process explained).

SESSION 2. Small groups. (Simultaneously, three small groups in separate sessions considered question "A" using NGT process).

BREAK (Tabulations of first session and results prepared on flip chart sheets for later display).

SESSION 3. Small groups. (Three small groups independently considered question "B" using NGT process).

BREAK (Tabulations of second session and results prepared on flip chart sheets for later display).

SESSION 4. General session, of all groups combined. (For question "A", NGT steps, 3, 4, and 5 are repeated).

SESSION 5. General session of all groups. (For question "B", NGT steps 3, 4, and 5 are repeated).

Each question required one hour and fifteen minutes to address using the following modified NGT process:

STEP 1. Generation of ideas in writing. (3 by 5 cards were used for each participant and each was requested to put their name and industry affiliation). The original thought for identifying industry affiliation was to compare information needs between segments. This was later determined inappropriate due to the small number involved.**

STEP 2. Round robin recording. (The facilitator asked each participant around the table, for one idea, writing each of these ideas on sheets of paper hung on the wall for all to see). Discussion was not allowed during this or the previous step.

** Maintaining a record of each participant's ideas and their subsequent individual rankings may offer considerable opportunities in needs segmentation analysis.

STEP 3.   Elimination of duplications. (This is the first step where group interaction was allowed. The facilitator asked the groups to eliminate duplicate ideas/statements. Both owners of the ideas had to agree.

STEP 4.   Combination of similar ideas/statements. (Again the owners of the statements had to agree that the statements were similar enough to combine. They also had to agree that the combined or newly written statement contained the essence of their intended ideas. The discussion in this step allowed for some clarification normally employed in Step 3 of the regular NGT process).

STEP 5.   Ranking of Priorities. (3 by 5 cards with numbers 1 through 7 were distributed and everyone secretly rank-ordered the top seven items. Another advantage of using the 3 by 5 cards is that every idea/statement is preserved. In this application it was important that all information needs be recorded despite their final position as viewed by the group as a whole).

Outcomes. The major outcome of the workshop was a list of priorities developed by consensus of the top ten Texas tourism information needs (for question "A" and question "B" as perceived by the 32 representatives participating in the workshop. Participants were very positive toward the NGT process. As can be noted from the needs identified, the diverse nature of the group seemed to produce more generic need statements. The more discrete needs identified in small groups tended to be incorporated or lost as the process worked. This is not viewed as a disadvantage given the goal of the workshop. The fact that all ideas generated at each phase were preserved in the final report of the workshop insured that all ideas generated were preserved.

10.4.2 Application #2

Goal. The goal of the second application of NGT was: "To produce a list of prioritized information needs of Texas commercial hunting enterprises." As in the first application, the NGT process was externally imposed by the Texas Tourism and Recreation Information Program as a means of identifying the information needs of target clientele for education and research programs. Again, however, the project was enthusiastically embraced by the owners and managers of the hunting enterprises.

Nature of Group. Participants represented two major geographic areas in Texas (the Rio Grande Plains and the Edwards Plateau) where a long tradition of commercial whitetail deer hunting exists. Participants were selected on the basis of meeting the following criteria:

1. They were operating hunting programs as a business enterprise.

2. They were the decision-makers for not only the hunting enterprise, but the ranches on which hunting took place.

3. They expressed the desire to improve the management of their enterprise through the use of more appropriate information.

The original list of owners/managers was selected by referral utilizing County Extension Agents, Extension specialists and university researchers knowledgeable of Texas hunting enterprises. From a list of over 60 landowners and managers, 30 agreed to participate in the workshop. Of that number 23, actually participated.

Preparation. The following two questions were formulated to generate the desired outcomes:

A. What kinds of information do you need in order to make better, more informed routine business decisions about your hunting enterprise and how it relates to ranch operations?

B. What kinds of strategic information do you need to help you get from where you are with your hunting enterprise, to where you want to be 3 to 5 years from now?

Each participant received the questions, an agenda and an explanation of the NGT process prior to the workshop. In addition to these materials, each participant received a two page questionnaire designed to profile participants by enterprise characteristics. This was a preparation step not utilized in the first application. Training was conducted as described in the first application.

Description of the Process. This workshop required approximately six hours to complete utilizing most of the same procedures used in Application #1. No changes were made in the NGT steps used in Application #1. One variation was made in the final general session after noticing some overlap in short and long term information needs. Instead of considering the small group lists for questions "A" and "B" separately, the small group results were combined. In other words, the top needs identified for question "A" and the top needs identified for question "B" were combined for the final general session. The last three steps of NGT were performed on the combined lists. The reasoning of facilitators was that the final result would show a priority relationship between short and long term information needs, which did occur.

Outcomes. The result of the second NGT workshop was a list of 34 information needs, ranked according to printing, of Texas hunting enterprises.

10.4.3 Application #3

Goal. The goal of the third NGT application was: "to determine those areas where 'collective action' can produce or contribute to Clear Lake area business prosperity." Unlike the first two NGT applications, this project was requested by the Clear Lake Boating and Recreation Council, a year-old organization of businesses and organizations concerned with enhancing boating and recreation business conditions. The perception of the leadership was that they had not been as effective as they wanted in getting the organization functioning. To this end, there were two sub-goals for the NGT process:

A.   To develop a priority list of factors which influence the prosperity
     of tourism related businesses in the Clear Lake area and designation
     of those which can be controlled through some "collective" effort.

B.   To develop an action plan which:
     1. Identifies tasks required to accomplish "collective" goals.
     2. Identifies who should be involved in each action program.
     3. Identifies information and resource needs necessary to
        accomplish desired ends.
     4. Identifies evaluation criteria to measure impact of
        initiatives/outcomes.

As can be noted, this application sought to go beyond information needs assessment to produce goals and action steps.

Nature of Group. Twenty-eight business and organizational leaders of the Clear Lake group received invitations to the workshop. All correspondence was issued through a local businessman designated as the coordinator for the group. Seventeen participated in the workshop.

Preparation. Because the group desired specific goals and action steps as outcomes, pre-workshop instructions were more detailed in order to inform participants of expected results. In addition to specifying the products which they expected, participants were given the following instructions and question:

> In preparation for the workshop, you should think about the factors which influence the prosperity of your business. The assumption is that as a business/owner in the Clear Lake area, you will manage the resources of the firm to accomplish your desired level of prosperity. There are also many factors outside the firm or business which influence your goals. The focus for this workshop are those factors outside the firm.

> Please respond in writing to the following question and bring to the workshop.

> QUESTION. What factors outside the firm do you feel are most influential in determining the prosperity of your business?

> Additional instructions requested they consider action steps, who should be involved, information needs, resources required and measures of success or failure.

Description of the Process. Six hours with an overnight break were required for the workshop. The break allowed time for participants to contemplate the prioritized needs generated the first day utilizing NGT. The 17 participants worked in two groups to produce a prioritized list of the top 10 factors influencing their business prosperity. Unlike the first two applications of NGT reported, another step was added to the general session for this group. The steps and sessions sequenced the same through the first voting round where items were ranked by the degree of "influence" on the business. The added step was to ask the group to rank the items a second time using the criteria of "control." In other words, they were asked to rank the factors based on their ability as a group to exert "control" over the factor.

A matrix was developed, that evening, to illustrate the combined weighted results of "influence" and "control" criteria. Figure 3 illustrates the resulting position of factors after combining the group's weighting. The top five factors based on these criteria were the focus for the next steps. To summarize, the following sequence represents those outlined in Application #1, plus those added in this application to produce the desired outcomes:

SESSION 1. Orientation.

SESSION 2. Small Groups.

SESSION 3. Omitted because there was only one question.

SESSION 4. General Session. (Items prioritized based on "influence" and again based on "control").

BREAK ---- Overnight Break. (Matrix developed to illustrate the combined weighted results of "influence" and "control").

SESSION 5. General Session. (Displayed matrix and discussion of top five factors. Instructions given for subsequent steps).

SESSION 6. Small Groups. Concurrent groups considered the top five factor and formulated response to the following for each:

* Determine goal for each of the top five factors.
* Determine measures for each goal.
* Determine time frame of goal accomplishment.
* Determine steps to achieve goal.

SESSION 7. General Session. Results of small groups displayed and consensus achieved on goals, measurements, time frame and steps.

SESSION 8. General Session. Participant evaluation of workshop.

The added steps provided opportunity for clarification. Spontaneous group interaction was used in Steps 6 & 7. Formal participant evaluation of the workshop to meet their expectations was added as a feedback mechanism (see Figure 10.3).

Outcomes. The outcome of the workshop was the development of four specific goals which participants felt they as a "collectivity" could control and achieve. A fifth goal, mentioned above was combined with another goal during the clarification process. Fourteen participants completed evaluations (three were not present for the last session). All agreed that the process met or exceeded their expectations. Six (43%) said it exceeded their expectations. In response to the question, "What was the single most important experience for you at the workshop?", 100% had positive comments about their ability to focus on common goals, work together or some aspect of agreement.

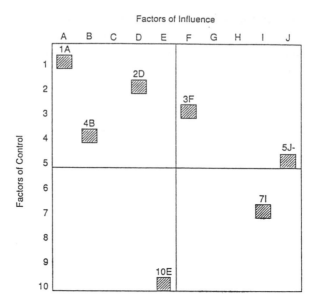

Figure 10.3   MATRIX OF HIGHEST RANKING FACTORS INFLUENCING
CLEAR LAKE BUSINESS PROSPERITY, RANKED BY
RELATIVE INFLUENCE AND DEGREE OF CONTROL.

10.4.4 Application #4

Goal. The fourth application of NGT was requested by the Galveston County
Council of Chambers of Commerce. The goal was: "to prioritize tourism objectives
and to produce an action program for their accomplishment." Unlike previous
application projects, the Council had already produced a tourism goal and a series
of specific objectives. The tourism goal and objectives had been produced as a part
of the County's Overall Economic Development Program (OEDP). The problem
perceived by the Council was the chronic lack of action taken on the specific
objectives.

Nature of Group. The Council of Chambers is comprised of representatives from
the communities and chambers of commerce of Galveston County. The chairman of
the Council was also the chair of the Overall Economic Development Committee
that produced the OEDP. The chair was responsible for invitations to the proposed
workshop.

Preparation. Since the overall goal and specific objectives had been developed,
formulation of question(s) to facilitate the process were not required. It had been
approximately six months since the adoption of the OEDP by the County
Commissioners Court and at least one significant piece of state legislation

authorizing local option voting on parimutuel betting. During that time the county had adopted the option. In view of this change and others that might not have been as apparent, invitees to the workshop were given the option of adding to the original ten OEDP tourism objectives.

Description of the Process. Thirty-four representatives of the Council were present for the workshop. Although invitees were requested to phone in their additional objectives, only one response was received. Given this situation, the first order of business after the orientation session was to solicit other objectives. Six additional objectives were generated in this fashion. It should be noted that more spontaneous discussion occurred in this process because of the considerable prior effort that had gone into the development of objectives for the OEDP.

The next step was to set a priority of the objectives based on two criteria:

A. By those objectives that will yield the greatest immediate impact on achieving the overall tourism goal.

B. By the ability of the group (including their individual influence) to control and accomplish the objective.

Each participant was given a 3 by 5 card and asked to rank the objectives by "impact." After collecting these cards, a second was distributed for ranking of objectives by "control." The variation in criteria used in this application was based on the need to focus on the goal which had already been articulated in the OEDP.

A matrix was developed to illustrate the combined weighing criteria on the objectives. Unlike Application #3, participants were given an opportunity to see the results prior to adjournment for the evening. The group was asked to consider how the top three objectives could be accomplished, including who should be responsible, how to measure, and appropriate time frame for accomplishment.

The second day began with a discussion of the prioritized objectives. As a result of the discussion, the group realized that the top three priority objectives contained a mixture of goals and objectives. With further discussion focusing on means of accomplishment, seven objectives which the group felt were more specific were articulated.

Further discussion of the seven objectives produced a consensus that the objectives could be grouped into two larger categories, one dealing with information and the other dealing with an institutional framework.

208

Two small groups were again formed to develop specific action steps for the two areas. A final general session combined the findings of the two groups.

Outcomes. The result of the fourth NGT application was the formulation of three recommendations:

1. Appoint tourism committee of Council of Chambers. (This recommendation resulted from the realization that the Council itself had no internal organizational structure to deal specifically with tourism).

2. The tourism committee then should develop a scope of work for developing a tourism information program. Elements of this program may include an inventory of all existing attractions and events, establishment of a central information distribution office/ communication point, development of strategic visitor information centers (possibly including existing convenience stores, and the like), development of a county-wide loop/slogan, a system of signage, etc.

3. The committee should develop a recommendation for the County Commissioners Court to create a paid tourism coordinator's position for the county.

The more interesting result of this final application of NGT was the significant departure from the original objectives articulated in the OEDP and produced in the first session. The fact that participants had apparently little sense of ownership in the OEDP objectives may have contributed to this phenomenon. In fact, most of the participants were not involved in the development of the tourism objectives in the OEDP. This might point out one of the limitations of the NGT process by illustrating that the product of NGT is a product of the participants involved. The OEDP objectives were not the product of the participants in the workshop, therefore the outcome was a different set of objectives. The implication is that NGT applications which purport to represent communities or diverse collectivities should be careful to include representation from all segments potentially impacted by outcomes.

10.5 Conclusions

The application of nominal group technique (NGT) by the Texas Tourism and Recreation Information Program was initiated to more clearly identify the information needs of targeted clientele. Each application of NGT during the last two years revealed its usefulness in accomplishing a variety of group objectives. A comparison of the four applications is shown in Table 10.2. The process was not initiated to test the efficacy of NGT, thus this reporting lacks the rigor that might have been imposed to control and document the process.

This paper is an attempt to look back on what was considered to be an extremely valuable application to tourism of a very useful group interactive tool. The fact that many tourism situations require collective action to deal with the external environment of the firm give sufficient justification to the need for group processes which are relatively free of bias from the facilitators and which can generate collective agreement in a non-threatening environment. Based on the experience of the authors, NGT provides that means.

Notable benefits of NGT observed in these applications include:

* It produces a feeling of personal involvement among all participants because each makes a contribution in generating ideas and in the final determination of priorities. This can be particularly meaningful in tourism when individuals of differing skills and positions come together to have input for collective action.

* The process generates many ideas in a short period of time. The small group idea generation sessions generated 40 to 50 items per group in the applications using this step (1, 2, and 3).

* The process facilitates obtaining input from people with different backgrounds and communication skills. While some of the groups in this reporting could be said to be more homogeneous than others, all could be generally characterized as having a variety of motives for being involved.

* NGT controls individual (hidden) agendas if organized and controlled as designed. On more than one occasion in these projects, it was apparent that some individuals were attempting to exert control on the outcome. In these situations it was observed that those individuals were satisfied despite their inability to control the process. One might conclude that the process is tragically democratic. It thwarts leadership, dominance by the better informed, or other biases that might be imposed in group inter-active situations.

TABLE 10.2   COMPARISON OF NOMINAL GROUP TECHNIQUE APPLICATIONS TO DETERMINE TOURISM INFORMATION NEEDS AND DEVELOP TOURISM GOALS FOR TTRIP CLIENTELE.

Application	How initiated[1]	Goal of NGT process	Nature of group (# of participants)	Degree of group homogeneity[2]	Required time for NGT process	Variations from NGT	Goals desired as outcome	Action plan desired as outcome
#1 Rockport Oct. 17, 1986	External to group	To identify/ prioritize information needs	19 segments of tourism industry (32)	Less	5½ hrs.	* Less discussion * Added 2 steps	No	No
#2 San Antonio Mar. 6, 1987	External to group	To identify/ prioritize information needs	Commercial hunting enterprise owners/managers (23)	More	6 hrs.	* Less discussion * Added 2 steps	No	No
#3 Clear Lake Nov. 17-18, 1987	Internal	To produce prioritized goals & action plan	Business and organizations of Clear Lake area (17)	More	6 hrs. (overnight)	* Added 2 steps * 2 criteria in voting * developed matrix	Yes	Yes
#4 Galveston County Jan. 26-27, 1988	Internal	To prioritize existing objectives and develop action plan	Galveston Council of Chambers Representatives (34)	Less	6 hrs. (overnight)	* 2 criteria in voting * developed matrix	N/A	Yes

[1] Refers to whether the group requested (internal) assistance or if NGT was externally initiated.
[2] Subjective evaluation of the relationship among group members.

The advantages of applying NGT seem very promising to tourism situations which require group conclusions. The following observations should be considered as one prepares to apply this tool:

*   NGT requires skilled group facilitators. The need for control and flexibility was required in the reported projects. The most notable need for being flexible was Application #4 where the outcome was significantly different than first anticipated. In every case, some modifications were required to accommodate the dynamics of the varied groups. Control was also required to avoid domination by individuals with dominating personalities.

*   The NGT step requiring the combining of similar statements can produce frustration if the group has difficulty with unclear working and overlapping concepts.

*   Clarification of ideas is another area where the experience of these reportings felt a possible weakness. This step requires adequate time and time constraints, as was generally the case, and may not allow adequate discussion.

*   Large groups are more difficult to control. The general sessions demonstrated the value of small groups. The larger the group the more difficult it seemed to insure uniform input from all participants.

*   A most important variable was the requirement for organization and attention to the smallest procedural detail. Much planning went into the preparation of questions, support materials, facilities, and support staff training.

Despite the required organization, preparation and need for skilled leadership, the NGT process is viewed as a most valuable tool.

A final observation should be made about the almost universal acceptance and feeling of value among participants. Participants seemed to generally reach a point where they realize the value of the process in identifying common areas of interest and benefit. Unfortunately, only participants in the third application were given opportunity to formally evaluate the process. Therefore, the observation that all felt the process valuable is based on the subjective evaluation of facilitators and their personal interactions with participants.

REFERENCES

Bartunek, Jean M. and J. Kenneth Murningham, 1984. The Nominal Group Technique: Expanding the Basic Procedure and Underlying Assumptions. Group and Organization Studies 9(3):417-432.

Butler, Lorna M. and Robert E. Howell, 1980. Coping With Growth: Community Needs Assessment Techniques. Western Rural Development Center, WREP 44, Corvallis, Oregon.

Delbecq, Andre, Andrew Van de Ven and David Gustafson, 1975. Group Techniques for Program Planning: A Guide to Nominal Group and Delphi Processes. Scott, Foresman and Co., Glenview, Illinois.

Erffmeyer, Robert C. and Irving M. Lane, 1984. Quality and Acceptance of an Evaluative Task: The Effects of Four Group Decision-Making Formats. Group and Organizational Studies 9(4):509-529.

Gill, Sandra L. and Andre L. Delbecq, 1982. Nominal Group Technique NGT. In: Group Planning and Problem-Solving Methods in Engineering Management, edited by Shirley A. Olsen, John Wiley and Sons, New York.

Hegedus, D.M. and R.V. Rasmussen, 1986. Task Effectiveness and Interaction Process of a Modified Nominal Group in Solving an Evaluation Problem. Journal of Management 12(4):545-560.

Lundberg, Craig C. and Alan M. Glassman, 1983. The Informant Panel: A Retrospective Methodology for Guiding Organizational Change. Group and Organizational Studies 8(2):249-264.

Watt, Carson, 1986. Texas Tourism Information Needs Assessment Utilizing Nominal Group Technique (NGT) with Selected Industry Representatives. Texas Agricultural Extension Service, Recreation and Parks Project Report, College Station, Texas.

Watt, Carson, 1987. Tourism and Recreation Information Program: Information Needs Assessment Project 2 -- Commercial Hunting Enterprise Operators. Texas Agricultural Extension Service, Recreation and Parks Project Report, College Station, Texas.

CHAPTER 11

Socio-cultural and environmental aspects of international tourism*

by
David L. Edgell, Sr., PhD
Senior Executive Director
United States Travel and Tourism Administration
and Adjunct Professor
The George Washington University

"A guest never forgets the host who had treated him kindly",
Homer: "The Odyssey", 9th Century B.C.

One of the most important motivations for travel is interest in the culture of other people in other countries. Socio-cultural and environmental aspects of international tourism relate to the way people learn about each other's way of life, thoughts and interactions with the environment. When a visitor has a positive cultural experience in a pleasant environment he won't forget "...the host who had treated him kindly."

There are representative expressions of people and places that provide powerful attractions for travel. People are interested in languages, expressions, cultures, and environments that are different from their own. Culture includes art, music. painting, sculpture, architecture as well as special events, celebrations and festivals. Or the cultural attraction may be food, drink, entertainment or other special forms of hospitality. The environment might be built with significant cultural, historical or hospitality orientation, or it might be a natural landscape, a pleasant seashore, a magnificent mountain, or lovely forest or it simply might be the social interactions of human beings. The concern herein is to recognize that such experiences may be positive or in some circumstances negative and to recognize the need for policy guidance to insure that the future growth of tourism will allow for balanced tourism experience.

The current rapid growth and development of tourism puts a special pressure on socio-cultural and environmental planning. All too frequently there are complaints that tourism generates pollution, crowding, crime, prostitution, and the corruption of the language, culture and customs of some local populations. This chapter will

---

* Reprinted with permission of the author from "Charting a Course for International Tourism in the Nineties: An Agenda for Managers and Executives". U.S. Department of Commerce, Washington, D.C., February, 1990.

explore some negative and positive aspects of socio-cultural and environmental issues in tourism.

11.1   Socio-cultural problems

It is clear that tourism has, in a short time-frame, led to a closer association and mingling of people of different races, creeds, religions and cultures. However, there is a growing concern that mass international tourism may have a detrimental impact on local cultures and customs or that a local area will distort its local festivals and ceremonies to "stage" spectacles for the benefit of international visitors. Thus, to some people tourism leads to the disappearance of traditional human environments and replaces them with towers of artificial concrete, ideas, ethics, morals and in effect, threatens the whole fabric of tradition and nature.[1] In addition, there is concern that as such distortions arise, the host and guest become a part of separate worlds leading to even greater prejudices and misunderstandings. For example, in the book *The Golden Hordes*, Turner and Ash contend that modern day tourism is a form of cultural imperialism, an unending pursuit of fun, sun, and sex by the golden hordes of pleasure-seekers who are damaging local cultures and polluting the world in their quest. In brief, the authors feel that a lot of "tourism" damages the local culture of the host country, perverts the traditional social values, encourages prostitution and hustling among "the natives", and usually results in a proliferation of fabricated tourist-oriented cultural performances and the sale of cheap souvenirs masquerading as local "arts and crafts".[2]

In other articles it is frequently mentioned that the attitude of the local population toward tourists is sometimes hostile due to the tourist's outlandish requirements for accommodations and services or due to the unreasonable wants and demands of limited number of rude and arrogant visitors. Other problems cited include the fact that in many of the developing countries there is foreign ownership and management of tourist facilities which may create the feeling that indigenous people perform only menial tasks. Tourism may be regarded as a threat to the indigenous culture and mores. There is often a perception that tourists are mainly responsible for the deterioration in standards of local arts and crafts as efforts are made to expand output to meet the tourist's demands. And, not infrequently, resort development has resulted in local people being denied access to their own beaches. All these factors can give rise to serious problems in the perception of tourists; and sometimes to demands for limitations on the flow of visitors.

11.2   Avoiding socio-cultural problems

While certainly there are numerous incidents where tourism does impact negatively on an area, this does not have to happen. A carefully planned, well-organized tourist destination can benefit the residents through exposure to a variety of ideas, people, languages and other cultural traits. It can add to the richness of the resident"s experience by stimulating an interest in the area's history through restoration and preservation of historical sites. For example, some of the cultural richness in the U.S. black communities is being revived as potential for tourism development.[3] The revitalization, for example, of Harlem has made that community a well-known tourist destination abroad. The myths, realities, folklore, and legacies of Harlem, New York, are now known around the world. It is increasingly being recognized both domestically and internationally for its rich cultural heritage, landmarks, museums, churches, parks, architectural structures, and varied night life.[4]

Organized cultural tourism development can provide opportunities for local people to learn more about themselves, thus increasing feelings of pride in their heritage and a heightened perception of their own self-worth. For example, residents of Mexico City speak with great pride about their Ballet de Folklorico, their National Museum of Anthropology and their Palace of Fine Arts. The Venezuelans speaks longingly and affectionately about "La Feria de San Sebastian", a great festive event with cultural and other exciting celebrations which not only draw Venezuelan participation but also includes foreign interest and visitation.

Even a highly localized heritage event such as, for example, "Cody Days", a festive occasion whereby the residents of Leavenworth, Kansas celebrate their historical link to William F. Cody--better known as "Buffalo Bill"--can be a positive cultural experience for non-residents as well. And the local celebration of "Potomac Days Parade" in Potomac, Maryland, is an event that has grown into an international festival whereby people of many different heritages ranging from Korea to Lithuania show off traditional clothes, food, and arts and crafts. These local celebrations started gradually but have grown into regular yearly celebrations that both residents and visitors look forward to.

Tourism can also contribute to cultural revival. There are numerous examples where the demand by tourists for local arts and crafts has heightened the interest and maintained the skills of local artisans and craftsmen by providing an audience and market for their art. In the United States, a number of Indian ceremonies and dances owe much of their continued existence to the fact that tourists were interested in them. This stimulated many local Indians to revive and teach to the new generations the meaning of such traditions. This preservation of cultural heritage, whether it be in local artifacts, historic sites or religious rites, forms the heritage of an area or country. It is often this very uniqueness which is the primary tourism attraction. It is a contribution to the quality of life of both the residents and tourists. But, it is often the tourists who provide the interest and economic means to preserve and maintain this cultural heritage.

The Manila Declaration (from the World Tourism Conference held in Manila in 1980) summarizes tourism's contribution to socio-cultural and environmental benefits as follows:

> *"The protection, enhancement and improvement of various components of man's environment are among the fundamental conditions for the harmonious development of tourism. Similarly, rational management of tourism may contribute to a large extent to protecting and developing the physical environment and the cultural heritage as well as improving the quality of life ... tourism brings people closer together and creates an awareness of the diversity of ways of life, traditions and aspirations ... "[5]*

In other words, the socio-cultural and environmental aspects of an area can enrich tourism in general, and provide different and unique opportunities for tourists to experience art, music, dance, food, literature, language, religion and history different from their own. At the same time, tourists bring to the local area socio-cultural traits from the homeland. This cross-cultural manifestation can have positive or negative results depending on the way tourism is handled in the receiving country.

For many years the OAS has been concerned with the need for enhancing the positive aspects of the socio-cultural and environmental impacts on tourism (particularly in the "Caribbean"). The numerous studies and reports by the OAS outline potential approaches for better understanding the socio-cultural and environmental aspects of tourism and for implementing programs to improve the socio-cultural and environmental impacts of tourism. In addition, the WTO has taken a strong interest in socio-cultural and environmental concerns for tourism. The WTO General Assembly, which met in Sofia, Bulgaria (September 1985), adopted a "Tourism Bill of Rights and Tourist Code" which contains several articles outlining socio-cultural concerns in tourism. Two brief excerpts suffice to illustrate the importance of this issue:

> *"... 1. Article VI. They (host communities) are also entitled to expect from tourists understanding and respect for their customs, religions and other elements of their cultures which are part of the human heritage. "*

> *2. Article VII. "The population constituting the host communities in place of transit and sojourn are invited to receive tourists with the greatest possible*

hospitality, courtesy and respect necessary for the development of harmonious human and social relations...".[6]

In planning for international visitors, the host country must understand the great socio-cultural variety in the backgrounds of the visitor as well as their reasons for the visit. Many people want the excitement of visiting new and different areas of the world, but at the same time they may be apprehensive about strange languages, customs and social structures. Salah Wahab in his book *Tourism Management* has one prescription for this paradox. He states: "The tourist country should be sufficiently different to be exciting and diversified, offering the tourist the novelty and escape he seeks, but sufficiently similar in comfort and security conditions to the tourist's own country to make him feel relaxed and at ease."[7] Certainly one of the challenges for planning a balanced tourism product is being able to take into account socio-cultural wants and needs of the visitor balanced with positive attributes for the host.

11.3   Some special cases of heritage tourism

Some communities seek to restore old buildings and similar edifices in an effort to maintain the historic preservation of the area and to draw visitors to participate in the local cultural heritages. In addition, there are examples, like Williamsburg, Virginia whereby the complete community is a replica of past history and culture. Still others, like the Polynesian Culture Center in Hawaii, seek to create visitor interest through the performance of dances and rituals of several different cultures in one place. While these efforts are often applauded, there are also many detractors who are critical and suggest that such portrayals as cited are artificial substitutes and often demean the society and culture on display.

An excellent example of a monumental effort to preserve a cultural, historical and environmental heritage through tourism is the five-nation regional project referred to as La Ruta Maya--the Maya Route. In an 81 page article in the October 1989 issue of *National Geographic*, author and editor Wilbur E. Garrett explains in great detail the opportunity that La Ruta Maya offers to "... increase environmentally oriented tourism and sustainable, nondestructive development to provide jobs and money to help pay for preservation". To get five countries, Mexico, Belize, Guatemala, Honduras, and El Salvador, to all agree to cooperate in this ambitious regional project provides a model for other parts of the world.[8]

Heritage tourism appears to be gaining widespread acceptance as both a part of the overall tourism effort and separately as a special attraction. Similar to other aspects of socio-cultural tourism, heritage tourism often creates a source of community pride which helps to ease resentment towards visitors and to prevent displacement of residents' business, particularly in downtown areas which often need economic revitalization and which often present an opportunity for cultural enrichment. The key is to balance the complaints of local residents about such problems as traffic congestion and lack of parking brought on by visitors, with the economic benefits and the potential for community pride in culture. It takes strong leadership and community support to overcome the obstacles and to explain the benefits that tourism can bestow on a local area. In summary, a well-planned effort can reap economic benefits, preserve buildings of historic significance, and create community pride in what the community offers to locals and outsiders alike.

11.4   Concerns for the environment

There have been numerous recent happenings which suggest that tourists and the environment are not very compatible. Some Tourists want souvenirs such as special corals, exotic rocks or lots of seashells. Others trample irreplaceable tundra or otherwise alter natural flora and fauna. Some people may want to chip off a piece of the Colosseum, or walk off with native artifacts or otherwise desecrate important man-made objects of historical and artistic importance.

The environment in which tourism interacts is broad in scope including not only land, air, water, flora and fauna, but also the man-made environment. The physical environment is just one facet of the surrounding with which the tourist comes in contact. As mentioned above, the tourist also must relate to socio-culture differences as well. In brief, the environment in its broad definition is what attracts the tourism in the first place. It may be the ecosystem, the wildlife, the rich archeological discoveries, the climate of the culture which the tourist may have read about, seen on the movie or television screen or been told about by a friend. The important note is that whatever the environment, it must be protected as an inheritance for future generations of visitors.

Recent commentaries paint a bleak picture for tourism's interaction with the environment. In an article entitled "Will There Be Any Nice Places Left?." a number of negative aspects of worldwide tourism are presented.[9] Polluted beaches, urban blight, eroded landscapes, and sprawling slumlike developments are mentioned as frequent sights in tourism areas. Many tourism developments are demeaning to local residents, overcrowded, noisy, conducive to traffic congestion, architecturally tasteless, and an overload on the infrastructure. Much of this kind of development in the past has been due to laissez-faire tourism policies and very little national, regional or local planning.

There has been increasing interest in recent years in the impact of tourism on the environment. More discussion has dwelled on environmental degradation caused by tourism than on positive aspects of tourism. The report of the 1973 European Travel Commission Conference on Tourism and Conservation stated rather forcefully both the positive and negative factors in the interdependent relationship between tourism and the environment:

> "First,... environment is the indispensable basis, the major attraction for tourism. Without an attractive environment there would be no tourism..."

> "Second,...the interests of tourism demand the protection of the scenic and historic heritage. The offer in the travel brochure must be genuine..."

> "In some countries, tourism...is seen by those concerned to protect the environment as their powerful ally. The desire to gain national income from tourism can impel governments to protect monuments or natural areas they might otherwise have neglected..."

> "Third, tourism can directly assist active conservation...can prompt men to contribute towards...conservation...of (famous places such as) Florence and...Venice." The entry fees of tourists help to maintain historic structures and parks... Tourist activity may provide new uses for old buildings...

> "And yet, despite these positive links, many conservationists feel that tourism can present a major threat to the environment...that countless hotels, roads and other facilities provided for the tourists ruin the beauties of the seacoast, disturb the peace of the country, and rob the mountains of their serene grandeur...streets choked with tourist traffic, and...squares and market places turned into parks for visitors."[10]

For Europe the relationship between tourism and the environment seemed to grow steadily worse throughout the 1970s. In an article from *The Washington Post*, August 1985, "Tourism Found Mixed Blessing by World Group," many of the negative features of tourism and the environment were highlighted based on an OECD survey. The article made the point that Europe had not been forceful in maintaining a balance between tourism and the environment.[11]

11.5   Some positive movement

More recently, people have become increasingly concerned about all aspects of pollution whether it be industrial, noise, people visual or tourism development. Even though most of the evidence suggests that the development of tourism infrastructure and facilities has generally caused less physical environmental damage than have timber and mineral extractions or industrial plants, there is concern for tourism development impacts on the environment in some areas. This concern has given rise to developmental constraints in an attempt to preserve the ecosystem and improve the quality of the environment. This recognition and protection has important benefits for the long-run health of the tourism industry. Without the protection of the scenic splendor that is often the very attraction to the tourist, the quality of tourism will deteriorate.

11.6   The challenge

The preceding paragraphs demonstrate that there does not have to be a negative connotation of tourism in relation to socio-cultural and environmental impacts. To the contrary, we can argue that there is a very natural interdependence of tourism and the culture and environment of a country. This concept was most eloquently addressed in a speech by the President of India, Giani Zail Singh, before the General Assembly of the World Tourism Organization in New Delhi on October 3, 1983. He said:

> "...Tourism can become a vehicle for the realization of man's highest aspirations in the quest for knowledge, education, understanding, acceptance and affirmation of the originality of cultures, and respect for the moral heritage of different peoples. I feel that it is these spiritual values of tourism that are significant... Tourism has also made it possible for nations to develop strategies for the conservation of natural and cultural heritage of mankind. Planning for economic growth and development must go hand in hand with the protection of environment, enhancement of cultural life, and maintenance of rich traditions which contribute so greatly to the quality of life and character of a nation. The rapid and sometimes alarming deterioration of environment due to pollution which is entirely man-made must be a matter for concern to all of us, who hold in trust on behalf of our peoples, the distinctive heritage of our respective countries..."[12]

To obtain this aspect of an improved "quality of life" is a challenge for tourism, particularly in the next ten years. It won't just happen. It will have to become an integral part of the policy and planning process for tourism development.

For example, if properly organized, tourism can provide an incentive for the protection of national parks, restoration of historical monuments and the preservation of cultural events. In many places in the world, the expenditures made by tourists are the economic means to protect the environment. For example, in a story in *U.S. News and World Report,* May 4, 1987, Bruce Wilson from the Center for Conservation Biology at Stanford University said that: "What has saved gorillas and cheetahs in Africa is the prospect of $500 million a year from safari-bound tourists."

ENDNOTES

[1] A good reference on cultural development and its relationship to tourism is contained in the Journal of the Mugla School of Business Administration, special issue, entitled "International Tourism Congress."

[2] Louis Turner and John Ash, The Golden Horde: International Tourism and the Pleasure Periphery, St. Martin's Press, 1976.

[3] For a more complete explanation see: "Cultural Richness in the U.S. Black Community Offers Great Potential for Tourism Development", by David L. Edgelll ad Bernetta J. Hayes. Business America, September 26, 1988, pp. 8-9.

[4] See "Tourism: An Economic Development Tool for Black and Minority Chambers of Commerce", by David L. Edgell. Business America, February 15, 1988, p. 5.

[5] Manila Declaration on World Tourism, World Tourism Conference, Manila, Philippines, September 27-October 10, 1980.

[6] World Tourism Organization meeting of the General Assembly, Sofia, Bulgaria, September 1985.

[7] Salah Wahab, Tourism Management, Tourism International Press, London, 1975.

[8] For a complete explanation of the "La Ruta Maya," see National Geographic, October 1989.

[9] "Will There Be Any Nice Places Left?", Next, September/October 1980, pp. 76-83.

[10] European Travel Commission, Tourism and Conservation: Working Together, London, 1974.

[11] The Washington Post, June 8, 1985.

[12] Text of a speech by the President of India, Giani Zail Singh, inaugurating the Fifth Session of the General Assembly of the World Tourism Organization in New Delhi on October 3, 1983.

[13] See also David L. Edgell, Sr., 1992. International Tourism Policy. Van Nostrand Reinhold, New York, 256 pages.

CHAPTER 12

Political and foreign policy implications of international tourism[*]

by
David L. Edgell, Sr., PhD
Senior Executive Director
United States Travel and Tourism Administration
and Adjunct Professor
The George Washington University

"Tourism is a simple continuation of politics by other means."
Jean-Maurice Thurot--Economia, May 1975.

12.1   The political economy of tourism

The political aspects of tourism are interwoven with its economic consequences. Thus, tourism is not only a "...continuation of politics..." but an integral part of the world's political economy. In short, tourism is, or can be, a tool used not only for economic, but for political means. For obvious economic reasons, most countries seek to generate a large volume of inbound tourism. As we learned, expenditures by foreign visitors add to national income and employment, and are a valuable source of foreign exchange earnings. Various measures are taken by governments to encourage foreigners to visit their respective territories. Promotion offices are established in key countries, bolstered by extensive advertising campaigns to attract tourists. Today more than 170 governments maintain travel promotion offices around the world. Most are located in the principal travel-generating countries of Western Europe, the United States, Canada, Mexico and Japan. Visas are issued freely for temporary visitors and other entry requirements are held to a minimum to avoid discouraging potential tourists (the United States is more restrictive than Western Europe, Canada, Mexico and Japan in its entry procedures). At home, governments seek to stimulate the construction of needed tourist facilities, access roads, communications, etc. Efforts are devoted to conserving areas of natural beauty, developing and maintaining resort areas and sightseeing attractions. Special events, entertainment, and cultural activities are often encouraged by local and

---

[*]   Reprinted with permission of the author from "Charting a Course for International Tourism in the Nineties: An Agenda for Managers and Executives". U.S. Department of Commerce, Washington, D.C., February, 1990.

national governments. Other services performed by governments are necessary to support tourism, such as police protection and crime control, and maintaining good health and sanitary conditions.

12.2   Tourism facilitation

A number of political, economic, and social factors influence the government actions and regulations affecting tourism. Travel bans are imposed from time to time for political reasons. It is not unusual, for example, for governments to prohibit travel of their citizens to war zones or to territories of hostile nations where the government has no means of protecting the life and property of its citizens. The U.S. State Department, for instance, through its Citizens Emergency Center, issues travel advisories to warn Americans considering going abroad about adverse conditions they might find in specific countries.[1] Special precautions also may be taken when outbreaks of contagious diseases occur in foreign countries and when these measures may result in discouraging or inconveniencing tourists. Also, some burdensome practices (exhaustive inspections of luggage, body searches) may be instituted for passenger safety and security, and to prevent smuggling.

Another concern of governments is immigration control. Nearly all countries strictly control the entrance of immigrants and enforce laws against illegal entrants. Of particular concern are the social pressures created by the need to care for jobless immigrants and opposition of the local labor force when jobs are scarce.

To even admit foreign visitors and to facilitate their travel within a nation's borders is a political action. Therefore, the way in which a nation's international tourism is approaches becomes a matter of its foreign policy, as well as a part of its economic and commercial policy.[2]

There are endless examples of the political and foreign policy implications of international tourism. The history of travel contains numerous references to international tourism with political overtones, ranging from Marco Polo's vivid descriptions of the political events in the orient to the uncertainty, lack of knowledge and myths associated with the "dark continent" of Africa prior to its exploration by the Europeans. Following are examples to illustrate the broad policy implications of international tourism in today's world.

Increased contacts between persons of different cultures can lead to increased knowledge and understanding which, in turn, can contribute to a relaxation of tensions between nations. For example, the Shanghai Communiqué, signed in 1972 by the United States and the People's Republic of China, noted, in part: "The effort to reduce tensions is served by improving communications between countries that have different ideologies, so as to lessen the risks of confrontation through accident, miscalculation or misunderstanding."[3]

There are numerous additional examples, but none more dramatic than what occurred in 1989 in Eastern Europe. Few individuals could have predicted the demise of the "Berlin Wall" in 1989 or the graphic television pictorials of soldiers cutting down the "Iron Curtain". Such changes are having profound effects upon East-West travel and will continue to do so right through the 1990s. The result will be a deeper understanding amongst peoples of the world, increased commerce and a greater step toward international cooperation. It will be sometime before we know just what the political implications of such human contact will eventually hold in store for us.

Even in a country like Cuba where tourism is sometimes referred to as a "Bourgeois" custom, international tourism is returning after a thirty-year hiatus. In a *The Washington Post* article dated August 3, 1987, entitled "Reviving the Allure of Cuba", it was pointed out that in Cuba's effort to earn hard currency it is now encouraging foreign visitors.[4] By 1985, Cuba saw 173,000 foreign visitors, mostly from Canada and Western Europe. According to an article "in Search of Tourists" in *Newsweek*, December 19, 1988, by then the number of foreign tourists grew to 211,464 by 1987.[5]

In a companion *Newsweek* article "White Sand, Blue Seas and Big Dreams", January 9, 1989, the problem for Cuba was cited as the need to dramatically increase tourism without upsetting "...the socialist life it is trying to build for its own citizens". Whatever the case, as more tourists visit Cuba, there will be political impacts of one kind or another.[6]

12.3   Tourism and foreign policy

The prospective economic benefits of tourism frequently influence the internal policies of governments. In some corners of our globe, inbound tourism is used to showcase the accomplishments of the government-or-party-in-power and to increase understanding abroad of the government's policies. Sometimes this approach is successful; sometimes, it backfires. The point is that tourism expands the horizon of the tourist and presents the host government or community with a unique opportunity to influence visitors from abroad. Alternatively, countries including the United States (U.S. Information Agency) sponsor numerous exchanges, cultural programs, lecture series and other events to make people of the world aware of country customs and standards of living. At the same time, a country must be made safe for tourism. Civil strife and disorders, such as those that have wrecked, for example, Northern Ireland and Lebanon, have a detrimental impact on tourism. In addition, the wave of terrorism in the mid-1980s weighed heavily on international tourism.

According to Jean-Maurice Thurot, tourists create an economic dependence by the host country on tourist-generating countries. This dependence can influence the foreign policy of the host country toward the generating country. This is especially true in nations needing foreign exchange, or hard currency, for economic development. Nations in the process of economic development need to buy key items, especially capital equipment and technology, from the industrial nations in order to speed their own growth. International tourism can be an engine of economic growth by providing an important source of foreign exchange. Most communist and less-developed nations need tourist dollars for economic growth. Government policy changes can accommodate tourism and thus decrease the need for merchandise exports.

12.4   Agreements beyond just tourism

Tourism has become imbedded in treaties and trade agreements designed and negotiated for largely non-tourism reasons. The most well-known international agreement containing tourism provisions is the human rights section of the 1975 Helsinki Accord, which was the Final Act of the Conference on Security and Cooperation in Europe. The better-known section of this accord deals with the rights of people to migrate freely, but in the tourism sections the thirty-five nations--including the United States and the Soviet Union--acknowledged that freer tourism is essential to the development of cooperation amongst nations. With specific reference to tourism, the signatories to the Accord, among other points, (a) expressed their intentions to "encourage increased tourism on both an individual and group basis." (b) recognized the desirability of carrying out "detailed studies on tourism," (c) agreed to "endeavor, where possible, to ensure that the development of tourism does not injure the artistic, historic and cultural heritage in their respective countries." (d) stated their intention "to facilitate wider travel by their citizens for personal or professional reasons," (e) agreed to "endeavor gradually to lower, where necessary, the fees for visas and official travel documents," (f) agreed to "increase, on the basis of appropriate agreements or arrangements, cooperation in the development of tourism, in particular, by considering bilaterally, possible ways to increase information relating to travel to other related questions of mutual interest, and (g) expressed their intention "to promote visits to their respective countries."[7]

But the hopes of this Conference and the potential of tourism as an agent of political reapproachment will be realized only through the efforts of governments, national tourist offices, and private industry. The record, until recently, has not been a promising one.

Tourism, as a foreign policy tool, was exercised in part by the U.S. Government in 1980, after the Soviet invasion of Afghanistan, when President Carter encouraged Americans and the U.S. Olympic Committee to boycott the 1980 Moscow Olympic Games. There was more than a 75 percent drop in American travel to the Soviet Union during the Olympic Year 1980. Tour operators abandoned travel to the Soviet Union and Eastern European markets in 1980-1981. There was a decided shift in tourism patterns. A somewhat similar situation occurred in 1984 when the East Bloc countries, spurred on by the Soviets (and with the exception of Romania), boycotted the Olympic Games in Los Angeles. To hopefully avoid future such happenings, the United States and the Soviet Union on September 15, 1985, signed an "Accord of Mutual Understanding and Cooperation in Sports" which is a step toward future cooperation in the Olympic Games.

Similarly, the United States, through the National Tourism Policy Act of 1981, put more emphasis on broader policy goals which often impact on foreign policy related goals. For example, under Title I of this Act, two of the twelve national tourism policy goals are to "... contribute to personal growth, health, education, and intercultural appreciation of the geography, history, and ethnicity of the United States", and "... encourage the free and welcome entry of individuals travelling to the United States, in order to enhance international understanding and goodwill..." Furthermore, Title II of the Act states that USTTA "... should consult with foreign governments on travel and tourism matters and, in accordance with applicable law, represent United States travel and tourism interests before international and intergovernmental meetings ..." This latter provision of the Act gives USTTA the authority to meet, negotiate and discuss a broad range of international tourism issues either bilaterally or multilaterally.[8]

12.5 International Tourism Policy for the United States

Because international tourism is such an important sector in almost every country, it is particularly appropriate that tourism services be considered in negotiations of bilateral agreements on trade in services or trade in tourism directly.[9] For the United States, tourism is increasingly being recognized as a tool for broader trade and policy goals for the country. A major step in this direction took place with the passage in 1981 of the National Tourism Policy Act.

The National Tourism Policy Act of 1981, 22 U.S.C. section 2122 et seq., in addition to the above, also directs the Secretary of Commerce to administer a comprehensive program to encourage travel to the United States, reduce barriers to

228

travel, and generally facilitate international travel. Section 2123 authorizes him, inter alia, to consult with foreign governments on travel and tourism matters, establish branches of official tourism offices in foreign countries, and assist in training and education on travel and tourism matters. The Act provides that the duties of the Secretary of Commerce in the administration of the Act may be delegated to the Under Secretary of Commerce for Travel and Tourism. The Under Secretary heads the U.S. Travel and Tourism Administration.

On a limited scale the United States Travel and Tourism Administration (USTTA) has fostered international tourism policies through international representations and bilateral negotiations with other countries. USTTA, in concert with the U.S. Department of State and other affected agencies, has negotiated tourism agreements with several countries. Fundamentally, tourism agreements are diplomatic arrangements prescribing reciprocal measures to reduce travel restrictions, facilitate two-way tourism and establishing the status of the parties' official travel promotion offices. However, each and every agreement has different and separate provisions depending upon the circumstances and concerns of the two countries. Most include the exchange of information and statistics as well as promoting greater understanding and goodwill through international tourism.

Since 1978, the United States has negotiated tourism agreements with eight countries: Mexico, Venezuela, Egypt, Philippines, Hungary, Yugoslavia, Poland, and Morocco. The most comprehensive agreements are with Mexico and Venezuela, both negotiated and signed in 1989. The first consultation and implementation of the U.S.-Mexico Tourism Agreement took place in Mexico City, November 21, 1989 between Secretary of Commerce Robert A. Mosbacher and Secretary of Tourism Carlos Hank Gonzalez.

These two agreements with Mexico and Venezuela differ from some other agreements on tourism by providing that each party, on a reciprocal basis, will accredit tourism promotion personnel of the other party as members of a diplomatic mission or consular post. This is in accordance with a movement by many countries (including the United States) to recognize that tourism activities of a national tourism office constitute a legitimate diplomatic and consular function within the meaning of Article 3 of the Vienna Convention on Diplomatic Relations. Such recognition gives stature to international tourism promotion and puts it on a more equal footing with other governmental activities.

The United States and Canada signed a Free Trade Agreement (FT) on January 2, 1988 which took effect on January 1, 1989 and which includes a tourism Annex. The FTA includes a "Service Chapter" which deals specifically with tourism. Under the FTA, the United States and Canada will accord "national treatment" to tourism services. National treatment means treatment no less favorable than the most favorable treatment accorded by a province or state to those residing within that province or state.[10] The FTA also specifies that annual consultations take place regarding the implementation of the tourism Annex of the FTA. The first such consultation took place the week of November 30, 1989 in Washington, D.C. The results of the consultation on tourism (other sectors consultations were also taking place) were summarized in a "Report by the Working Group on Tourism", issued in Washington D.C. on November 30, 1989.

12.6   Intergovernmental organizations

There are at least eight intergovernmental organizations identified as being involved with international policy relating to problems and issues in tourism. Three principal organizations are the Organization of American States (OAS). Organization for Economic Cooperation and Development (OECD) and the World Tourism Organization (WTO). Following is a very brief description of the involvement of these organizations with international tourism.

12.6.1 Organization of American States (OAS)

The OAS, headquartered in Washington, D.C., is currently composed of the following countries: Antigua and Barbuda, Argentina, the Bahamas, Barbados, Bolivia, Brazil, Canada, Chile, Colombia, Costa Rica, Cuba, Dominica, Dominican Republic, Ecuador, El Salvador, Grenada, Guatemala, Haiti, Honduras, Jamaica, Mexico, Nicaragua, Panama, Paraguay, Peru, St. Kitts and Nevis, Saint Lucia, Saint Vincent and the Grenadines, Surinam, Trinidad and Tobago, United States, Uruguay and Venezuela.

The OAS held its first Inter-American Travel Congress (meets every three years) in San Francisco in 1939. Much of the significant work of the OAS in tourism was accomplished through its Tourism Development Program formed in 1970. The main functions were to assist member tourism authorities in developing and promoting their respective tourism sectors; support member states' efforts to create appropriate conditions for increasing the flow of tourism to the region; provide broad policy advice on tourism issues and coordinate with international bodies on tourism matters. The principal policy bodies of the OAS tourism program are the Executive Committee (seven elected member nations which meet about once a year) and the Inter-American Travel Congress (which meets every three years and also includes invited observers from non-member nations). Some of the important tourism matters dealt with included financing mechanisms, facilitation, statistics and education and training.

The OAS, over the years, has viewed tourism as having broad policy implications beyond the narrow economic benefits so important to most of the countries. At a "special" meeting of the Inter-American Travel Congress in Rio de Janeiro, Brazil, on August 25, 1972, the OAS formulated the *Declaration of Rio de Janeiro*, an important document which relates tourism to some of the broader issues. This document was reinforced through the *Declaration of Caracas* at the XIII Inter-American Travel Congress, Caracas, Venezuela, September 24, 1977. In addition, the OAS has been an effective partner in the numerous activities underway to celebrate the 500th Anniversary of the arrival of Christopher Columbus in the Western Hemisphere in 1992. The planning for this event is taking place under the auspices of the "Quincentennial Commemoration of the Discovery of America: The Encounter of Two Worlds, and Opportunities for Tourism Promotion".

In a recent reorganization of the OAS, the tourism programs have been integrated into the OAS' Department of Regional Development.

12.6.2 *Organization for Economic Cooperation and Development (OECD)*

The OECD, headquartered in Paris, in a forum for consultations and discussions by most of the industrialized countries on a broad range of economic issues. Through various committees and working groups, the OECD conducts studies and negotiations to solve trade and related problems and to coordinate its policies for purposes of other international negotiations. The OECD Tourism Committee reviews problems in international travel and tourism among member countries and publishes statistics and policy changes in a yearly publication entitled *Tourism Policy and International Tourism in OECD Member Countries*. A milestone was achieved in November 1985, when the OECD Council adopted a new instrument on international tourism policy which set forth principles aimed towards facilitation of tourism for the 24 member countries. It is a major attempt at reducing restrictions on tourism and setting in motion a process of liberalization.

12.6.3 *World Tourism Organization (WTO)*

The only worldwide (109 member countries in 1989) tourism organization is the WTO, headquartered in Madrid. It was formally established on January 2, 1975. It provides a world clearinghouse for the collection, analysis and dissemination of technical tourism information. It offers national tourism administrations and organizations the machinery for a multinational approach to international discussions and negotiations on tourism matters.

It includes more than 150 affiliate members (important private sector companies) interested in international dialogue and implementation of worldwide conferences, seminars and other means for focusing on important tourism development issues and policies. WTO is also an implementing agency on technical assistance in tourism for the United Nations Development Program. WTO has an important Facilitation Committee whose aims are to propose measures to simplify entry and exit formalities, report on existing governmental requirements or practices which may impede the development of international travel, and develop a set of standards and recommended practices for a draft convention to facilitate travel and tourist stays through passport, visa and health and exchange control measures.

The WTO has conveyed its broad concerns for all aspects of tourism through a number of special documents. The most popular and most often cited document is the *Manila Declaration on World Tourism* prepared during "The World Tourism Conference" held at Manila, Philippines from September 27-October 10, 1980 as sponsored by the WTO. Another important document is the *Tourism Bill of Rights and Tourist Code*. After several years of consideration and negotiations, the *Tourism Bill of Rights and Tourist Code* was adopted by the Sixth General Assembly of WTO in Sofia, Bulgaria, in September 1985. The most recent document of note is *The Hague Declaration on Tourism*, which was adopted during the "Inter-Parliamentary Conference on Tourism", April 10-14, 1989, jointly sponsored by the Inter-Parliamentary Union and the WTO.

While considerable progress has been made in utilizing tourism as an international policy tool for greater economic development and improved communication, cooperation, mutual understanding and goodwill, there is much that remains to be accomplished. Barriers to international tourism continue to exist and in some circumstances have increased. And, while it may be overly-optimistic to expect that the WTO's motto, "Tourism: passport to peace," will be shared by everyone, it is a step in the right direction. We do know that when peace prevails, tourism flourishes.[11]

12.7   International terrorism

A discussion of political and foreign policy implications of international tourism would not be complete in today's world without mention of international terrorism and its impact on tourism. Terrorism is not new; it is an age-old economic weapon. The history books are full of accounts of terrorism, hostage taking and kidnapping. The infamous period of piracy on the high seas with its plunder and violence was a time of terrorism and kidnapping which wreaked havoc on the maritime industry.

What's new in terrorism is its use to attain political ends and the global attention that media coverage of terrorist incidents focuses on political causes. Also new are some of the responses. For example, the U.S. attack on Lybia and various countries uses of military personnel to assault the hijackers while the plane is on the ground. The horror of these events and the highly publicized results were enough to cause many travellers to reconsider their vacation plans.

One of the terrorists' principal objectives is to exact a price, to inflict an economic penalty, to punish governments for public policies and political behavior; in short to make it prohibitively expensive for a government to carry out policies terrorists find unacceptable. By murdering innocent tourists, blowing up airliners and airline ticket offices, bombing airport terminals and railroad tunnels, and hijacking airliners, buses and cruiseships, terrorists seek to:

- frighten away tourists;
- deny governments, and populations, the commercial and economic benefits of tourism; and
- force governments to rethink and abandon specific policies.

The response by governments and the private sector to the impact of terrorism in tourism surpasses any prior attentions to security. The U.S. Federal Aviation Administration has increased its inspections of airports worldwide. Airport authorities have increased security systems within the airports. It will take a strong concerted effort and global cooperation if the "terrorism of the eighties" is to be avoided in the nineties.

ENDNOTES

[1] A good explanation of travel advisories is contained in Tour and Travel News, June 5, 1989.

[2] Howard F. Van Zandt, Ed., International Business Prospects 1977-1999, Bobbs-Merrill Educational Publishers, Indianapolis, 1978, pp. 171-177.

[3] Shanghai Communiqué (February 27, 1972); a joint statement issued at the conclusion of President Richard M. Nixon's visit to China.

[4] This article also contains an interesting commentary on various paradoxes facing Cuba which on the one hand is developing tourism for foreigners but doesn't allow Cubans to mix with tourists.

[5] The article suggests that tourism may replace sugar as Cuba's main source of hard currency. As tourism increases there is a greater need for new infrastructure and Cuba is now seeking joint venture arrangements with foreign companies (mostly with Spain so far).

[6] Tourism is already causing changes that somehow seem to clash with Cuba's ideologies. For example, this article mentions that Ernest Hemingway, an American, is frequently used as a promotional gimmick to attract tourists. There are presumably all kinds of "recreated" bars and restaurants advertising that Hemingway had been there or that it was his favorite place.

[7] The Conference on Security and Cooperation in Europe, Final Act (commonly referred to as the "Helsinki Accord") was signed by 35 nations on August 1, 1975, in Helsinki.

[8] Business America, May 28, 1984, P. 2.

[9] For further explanation of the provisions of the National Tourism Policy Act, see "Recent U.S. Tourism Policy Trends", International Journal of Tourism Management, Volume III, No. 2, June 1982, pp. 121-123.

[10] This section is based on a presentation made by Mindel De La Torre, U.S. Department of Commerce on May 9, 1989 at the combined seminar of the Tourism Policy Forum (The George Washington University) and the American Bar Associations Subcommittee on Travel and Tourism for the International Service Industries Committee held at The George Washington University. This author also made a presentation on the status of negotiations on the bilateral tourism agreements, with Mexico and Venezuela at this seminar. For a summary of the seminar by Milton Zall, see: "Tourism Policy Forum Brief", Volume 1, Number 3, The George Washington University, Washington, D.C., December 1989.

[11] Journal of Travel Research, XXII, Number 3, Winter 1984, pp. 15-16.

[12] See also: David L. Edgell Sr., 1992. International Tourism Policy. Van Nostrand Reinhold. New York, 256 pages.

LIST OF CONTRIBUTING AUTHORS

Adri Dietvorst is Chairman at the Center for Recreation Studies at the Agricultural University, Wageningen, The Netherlands. He is also Head of the Department of Outdoor Recreation and Tourism at the Winand Staring Centre, Wageningen, The Netherlands.

David L. Edgell is Senior Executive Director of the U.S. Travel and Tourism Administration, Department of Commerce, Washington D.C., U.S.A. He is also Adjunct Professor of the George Washington University in Washington D.C., U.S.A.

Donald F. Holecek is Director of the Michigan Travel, Tourism and Recreation Resource Center at Michigan State University East Lansing, Michigan, U.S.A.

Janos Jacsman is Doctor at the Institute for National, Regional and Local Planning ETH (Swiss Federal Institute of Technology), Zürich, Switzerland.

Hubertus Mittmann is Regional Landscape Architect at the United States Department of Agriculture, Forest Service, Rocky Mountain Region, in Lakewood, Colorado, U.S.A.

Joe Porter is Principal in Design Workshop in Denver, Colorado, U.S.A.

René Schilter is from the Institute for National, Regional and Local Planning ETH (Swiss Federal Institute of Technology), Zürich, Switzerland.

Willy Schmid is Chairman at the Institute for National, Regional and Local Planning ETH (Swiss Federal Institute of Technology), Zürich, Switzerland.

James Stribling is associate Professor and Extension Specialist at Texas Agricultural Extension Service, Department of Recreation, Park and Tourism Sciences, Texas A. and M. University, College Station, Texas, U.S.A.

Jean Tarlet is Doctor in Geography and Planning at the Service Watermanagement and Land Use Planning in the Provence Region, Aix en Provence, France.

Pat Taylor is President of Taylor and associates, Dallas, U.S.A. He is also Director of the Program in Landscape Architecture at the University of Texas and Adjunct associate Professor of the Department of Recreation, Park and Tourism Sciences, Texas A. and M. University. Together with Hubert van Lier, he is founder of ISOMUL

236

Hubert van Lier is Chairman of the Department of Physical Planning and Rural Development and Head of the section of Land and Water Design of the Agricultural University, Wageningen, The Netherlands. Together with Pat Taylor, he is founder and currently also chairman of ISOMUL.

Turgut Var is Professor at the Department of Recreation, Park and Tourism Sciences, Texas A. and M. University, College Station, Texas, U.S.A.

Carson Watt is Interim Head of the Department of Recreation, Park and Tourism Sciences, Texas A. and M. University, College Station, texas, U.S.A.

INDEX

240